D0553826

Protective Clothing Systems and Materials

OCCUPATIONAL SAFETY AND HEALTH

A Series of Reference Books and Textbooks

Occupational Hazards · Safety · Health
Fire Protection · Security · Industrial Hygiene

1. Occupational Safety, Health, and Fire Index, *David E. Miller*
2. Crime Prevention Through Physical Security, *Walter M. Strobl*
3. Fire Loss Control, *Robert G. Planer*
4. MORT Safety Assurance Systems, *William G. Johnson*
5. Management of Hotel and Motel Security, *Harvey Burstein*
6. The Loss Rate Concept in Safety Engineering, *R. L. Browning*
7. Clinical Medicine for the Occupational Physician, *edited by Michael H. Alderman and Marshall J. Hanley*
8. Development and Control of Dust Explosions, *John Nagy and Harry C. Verakis*
9. Reducing the Carcinogenic Risks in Industry, *edited by Paul F. Deisler, Jr.*
10. Computer Systems for Occupational Safety and Health Management, *Charles W. Ross*
11. Practical Laser Safety, *D. C. Winburn*
12. Inhalation Toxicology: Research Methods, Applications, and Evaluation, *edited by Harry Salem*
13. Investigating Accidents with STEP, *Kingsley Hendrick and Ludwig Benner, Jr.*
14. Occupational Hearing Loss, *Robert Thayer Sataloff and Joseph Sataloff*
15. Practical Electrical Safety, *D. C. Winburn*
16. Fire: Fundamentals and Control, *Walter M. Haessler*
17. Biohazards Management Handbook, *Daniel F. Liberman and Judith Gordon*
18. Practical Laser Safety: Second Edition, Revised and Expanded, *D. C. Winburn*
19. Systematic Safety Training, *Kingsley Hendrick*
20. Cancer Risk Assessment: A Quantitative Approach, *Samuel C. Morris*
21. Man and Risks: Technological and Human Risk Prevention, *Annick Carnino, Jean-Louis Nicolet, and Jean-Claude Wanner*
22. Fire Loss Control: A Management Guide. Second Edition, Revised and Expanded, *Peter M. Bochnak*
23. Computer Systems for Occupational Safety and Health Management: Second Edition, Revised and Expanded, *Charles W. Ross*
24. Occupational Hearing Loss: Second Edition, Revised and Expanded, *Robert Thayer Sataloff and Joseph Sataloff*
25. Protective Clothing Systems and Materials, *edited by Mastura Raheel*

ADDITIONAL VOLUMES IN PREPARATION

Protective Clothing Systems and Materials

edited by

Mastura Raheel
University of Illinois at Urbana-Champaign
Urbana, Illinois

Marcel Dekker, Inc. **New York • Basel • Hong Kong**

Library of Congress Cataloging-in-Publication Data

Protective clothing systems and materials / edited by Mastura Raheel.
p. cm. -- (Occupational safety and health; 25)
Includes bibliographical references and index.
ISBN 0-8247-9118-5 (acid-free paper)
1. Protective clothing. I. Raheel, Mastura. II. Series: Occupational safety and health (Marcel Dekker, Inc.); 25.
T55.3.P75P76 1994
604.7--dc20 93-46029
CIP

The publisher offers discounts on this book when ordered in bulk quantities. For more information, write to Special Sales/Professional Marketing at the address below.

This book is printed on acid-free paper.

MARCEL DEKKER, INC.
270 Madison Avenue, New York, New York 10016

Current printing (last digit):
10 9 8 7 6 5 4 3 2 1

PRINTED IN THE UNITED STATES OF AMERICA

Preface

Interest in protective clothing has grown tremendously in the past decade due to greater awareness about worker safety and health in atypical or hazardous work environments. Government agencies such as the Occupational Safety and Health Administration (OSHA) and the Environmental Protection Agency (EPA) have issued guidelines, recommendations, and rules for worker safety in various occupational hazard situations, whereas industry and independent organizations have developed standards for personal protective equipment (PPE). At the same time, major strides have been made in the development and marketing of new high-performance materials (textiles, polymers, composites) for protective clothing. This trend is supported by an increasing demand for totally protective personal equipment such as fully encapsulated suits for use in hazardous waste disposal sites, when responding to chemical spills or fires, or in high-hazard industrial operations.

There are many occupations that require specific activities and different levels of protection. Hence, a large variety of protective clothing and equipment is currently available and new products are being introduced continuously. The success of an industrial hygiene, health, and safety program rests on the appropriate selection, use, and maintenance of personal protective equipment, as well as on training of employees in the proper use of the selected apparatus. This volume focuses on scientific and technical developments in this field, as well as on practical information regarding protective apparel systems vital to workers' safety in hazardous environments. The scope of hazardous conditions is limited to chemical, thermal, and microbiological environments.

Protective apparel systems are presented in three sections. The first deals with chemical protective clothing. It includes current literature on barrier properties and chemical resistance of various textile systems, refurbishing reusable contaminated clothing, and attitudes toward chemical protective clothing. The second section covers thermal protective clothing. It includes the theory of heat transfer from the body as well as clothing systems and focuses on heat- and fire-resistant textiles, and flame-retardant finishes used for protective clothing. The final section discusses microbiological-resistant fabrics and biological test methods for protective clothing.

This volume will be useful to various agencies responsible for specifying guidelines for industrial workers' safety and health. Industries that may require various levels of worker protection—and hence various types of protective clothing—include chemical, glass, and metal manufacturing; automobile; aerospace; electronic; and service industries such as health care, electric, gas, fire prevention, and waste cleanup. Military and Coast Guard personnel, as well as agriculturists (such as pesticide applicators), also will find relevant information. Academic researchers and professionals interested in protective apparel research, as well as manufacturers and distributors of protective clothing systems, should find this volume useful.

Mastura Raheel

Contents

Contributors

Elizabeth P. Easter, Ph.D. Associate Professor, Department of Interior Design, Merchandising and Textiles, University of Kentucky, Lexington, Kentucky

Norman W. Henry III, M.S. Senior Research Chemist, Central Research and Development Department, Haskell Laboratory, E. I. du Pont de Nemours & Company, Newark, Delaware

Charles J. Kim, Ph.D. Professor, Department of Clothing and Textiles, University of North Carolina at Greensboro, Greensboro, North Carolina

Joan Laughlin, Ph.D. Professor, Department of Textiles, Clothing, and Design; Associate Dean, Research, College of Human Resources and Family Sciences, University of Nebraska—Lincoln, Lincoln, Nebraska

Kenneth C. Parsons, Ph.D. Department of Human Sciences, Loughborough University of Technology, Loughborough, Leicestershire, United Kingdom

Theresa A. Perenich, Ph.D. Professor, Department of Textiles, Merchandising, and Interiors, The University of Georgia, Athens, Georgia

Mastura Raheel, Ph.D. Professor of Textile Science, Division of Consumer Sciences, University of Illinois at Urbana-Champaign, Urbana, Illinois

Margaret Rucker, Ph.D. Professor and Chair, Division of Textiles and Clothing, University of California at Davis, Davis, California

Tyrone L. Vigo, Ph.D. Lead Scientist, Textile Finishing Chemistry Unit, Southern Regional Research Center, Agricultural Research Service, U.S. Department of Agriculture, New Orleans, Louisiana

1

Protective Clothing: An Overview

MASTURA RAHEEL University of Illinois at Urbana-Champaign, Urbana, Illinois

I. INTRODUCTION

The growing concern regarding health and safety of workers in various sectors of the industry has generated regulations and standards, environmental and engineering controls, as well as tremendous research and development in the area of personal protective equipment (PPE). Personal protective equipment includes personal protective clothing and gear such as respirators, face masks, and other controls. The focus of this discussion, however, is on personal protective clothing.

All clothing is protective to some extent. It is the degree of protection from a specific hazard that is of major concern. The nature of a hazard to the health and safety of workers can be uniquely different. Nevertheless, workplace hazards can be grouped in the following general categories:

1. Chemical
2. Thermal
3. Mechanical
4. Nuclear (radiation)
5. Biological

Protective clothing challenges can often include combinations of categories, for example, chemical protective clothing (CPC) may be subject to heat, mechanical hazards (cut, tear, puncture), and complex chemical mixtures.

Clothing design, fitting problems, and lack of product uniformity during manufacture of protective clothing can cause additional complications for users [1].

Protective clothing and equipment is used in the chemical industry to prevent exposure to chemicals during production, distribution, storage, and use. These chemicals may be gases, liquids, or solids. The broad range of individuals who may be exposed occupationally to chemicals includes chemists, agriculturists, horticulturists, structural pesticide applicators, emergency response teams such as the coast guard and fire brigades, hazardous waste cleanup crews, off-shore petroleum workers, propellant handlers, and textile dyeing and finishing or paper printing industry workers. By the same token, the type of protective clothing used may range from everyday work clothing in conjunction with a pair of protective gloves to chemically resistant coveralls, gloves, face shields, and respirators to full-body encapsulating suits.

Protection from hot or cold environments requires protective clothing with the ability to insulate against conductive thermal transfer or provide protection against burn injuries. Fire fighters, foundry and glass manufacturing workers, and utility services, aerospace, and military personnel may encounter hazards associated with thermal environments.

It is not sufficient to rely solely on the fire- and heat-resistant properties of the materials used in protective clothing; the performance of the fabric assembly as it would exist in clothing, against various types of heat exposure, is also important [2–6]. For example, different materials might be recommended against convective (flames), conductive (molten metal splashes), and radiant (fire situations, metal and glass manufacturing) sources of heat.

Mechanical hazards are generally classified into two types: civilian hazards and military hazards. Civilian hazards may involve ballistic threats encountered in police and law enforcement work; ballistic threats related to personnel assassination, riots, hunting, and mine explosions; or threats from sharp objects encountered in packaging, glass, and lumber industries. Two types of soft armor are prepared from fabrics of high-strength, high-energy absorbing fibers such as Kevlar aramid. These are soft body armor and soft armor structures. Protective clothing or soft body armor is made of woven fabrics, nonwoven fabrics, reinforced composites, or combinations of these. Soft body armor items include bullet-resistant vests and military flak jackets to protect the torso of a human body, protective leggings for workers in the lumber industry to protect from chain saw mishaps, or protective cut-resistant gloves and aprons used in the butcheries and glass-manufacturing industries. Soft armor structures, which are used on structures to protect personnel or equipment, include blankets, curtains, and liners.

Shielding radiation from both x-rays and gamma rays is important for personnel in medical fields, nuclear power stations, and other inspection facilities where radiation is involved. Lead metal is known for its efficiency to block radiation. The major problem is how to make flexible protective clothing using the metal. In the past, shielding mats containing lead compounds such as lead sulfate and lead oxide have been used to produce aprons, partitions, etc. Recently, Toray Co. of Japan has produced melt spun fibers of lead metal, which can produce shielding mats with good radiation protection at lower mat thickness. The sheet exhibits excellent flexibility and ease of cutting and sewing. Thus, it is suitable for making work clothing (vests) for nuclear power station workers or x-ray shielding aprons, partitions, or curtains for use in the medical industry.

Protective clothing against biohazards is currently used in hospitals and other confined environments where transmission of diseases caused by microorganisms may occur. A recent ruling by OSHA [7] requires health care and other personnel to wear appropriate protective clothing when handling body fluids and/or contaminated laundry that may transmit insidious viruses such as hepatitis and human immunodeficiency virus (HIV). Physicochemical methods are employed to render clothing as barriers against microbial contamination and transmission through it. One method to impart barrier properties to garments, surgical gowns, drapes, and related materials is to coat them with impermeable coatings. Other methods include microencapsulation of antimicrobial agents within the fibers and controlled release of active antimicrobial agents.

The major focus of this overview is on chemical, thermal, mechanical, and biological protective clothing; hence, these aspects will be further discussed.

II. PROTECTION FROM CHEMICAL HAZARDS

The users of chemical protective clothing are faced with a difficult task of selecting appropriate clothing. Among the many factors that must be considered, the most important is the effectiveness of the CPC as a barrier to the chemicals of interest. Indeed, the style and construction, comfort factors, mode of use (one time or multiple use), and cost also have an impact on the selection of CPC.

In the mid-1970s, the Industrial Safety Equipment Association (ISEA), a protective clothing and equipment manufacturers trade organization, expressed concern about the lack of standards for industrial protective clothing. At the request of ISEA and with the agreement of several government groups, research organizations, and protective clothing users, in 1977 the American Society for Testing and Materials (ASTM) formed a Committee F-23 on Protective Clothing.

Initially, the focus of Committee F-23 was on chemical protective clothing only, but in 1980 the scope of Committee F-23 was expanded to include all aspects of materials and protective clothing end-use items for protection against occupational exposure to chemical and thermal hazards. Committee F-23 has developed several standards for chemical barrier and other performance testing of chemical protective clothing [8].

There is a need for information about the type and level of protection afforded by CPC against target chemicals and especially chemical mixtures. The information is vital for emergency response personnel such as fire fighters, coast guard emergency response teams, and chemical waste cleanup crews. Major strides have been made in the area of polymer materials development, barrier testing of materials (and end products), and development of standard testing methods.

A. Polymer Materials Development

The inherent chemical and physical properties of materials used in CPC are critical factors that must be considered early in the selection process for a particular application. Polymer materials used for CPC may range from natural cotton fiber and its blends with other fibers such as polyester used in ensembles frequently used by pesticide applicators, particularly in hot climate, to synthetic fibers and impermeable films used as nonwoven disposable materials or reusable materials. Some examples of disposable materials for coveralls are nonwoven polyethylene (PE) (Tyvek), or PE-coated Tyvek or Saran-laminated Tyvek (Saranex). Some reusable materials are butyl rubber, nitrile, neoprene, and chlorinated polyethylene suit and glove materials. In recent years, several high-technology, chemical-resistant materials have been developed, such as Teflon—a fluorinated-ethylene-propylene (FEP) material for face shields—and Viton—another fluorinated material developed by DuPont Co. for protective gloves and CPC specifically for protection against chlorinate hydrocarbons. The polymers can be used as single-layer garment materials or as laminates with other materials. Another high-performance material, ChemFab Teflon used in making Challenge 5100 and 5200 (a proprietary, aramid-reinforced fluoroelastoplastic composite), has been developed by Chemical Fabrics Corporation [9]. This material has exhibited a high level of chemical resistance when tested against a battery of 115 liquid chemicals at the U.S. Coast Guard research and development center [9]. Challenge 5100 also has undergone vapor permeation and physical integrity testing after exposure to ultraviolet (UV) light, ozone, as well as temperature changes.

Chemical resistance is evaluated by using ASTM "Test Method for Resistance of Protective Clothing Materials to Permeation by Liquids and Gases" (ASTM-F739-85) against a representative battery of chemicals specified in ASTM "Standard Guide for Test Chemicals To Evaluate Protective Clothing Materials" (ASTM-F1001-86). A 3-hr period is generally specified to assess the compatibility of test chemicals. The list of chemicals recommended in ASTM F1001-86 is presented in Table 1.

Table 1 Standard Chemicals Recommended for Evaluating Protective Clothing Materials

Chemical	Chemical class
Liquid	
Acetone	Ketone
Acetonitrile	Nitrile
Carbon disulfide	Sulfur organic compound
Dichloromethane	Chlorinated hydrocarbon
Diethyl amine	Amine
Dimethylformamide	Amide
Ethyl acetate	Ester
Hexane	Aliphatic hydrocarbon
Methanol	Alcohol
Nitrobenzene	Nitrogen organic compound
Sodium hydroxide (50%)	Inorganic base
Sulfuric acid (93.1%)	Inorganic acid
Tetrachloroethylene	Chlorinated hydrocarbon
Tetrahydrofuran	Heterocyclic ether
Toluene	Aromatic hydrocarbon
Gaseous	
Ammonia, anhydrous (99.9%)	Basic gas
1,3-Butadiene (99.0%)	Unsaturated hydrocarbon gas
Chlorine (99.5%)	Acid gas
Ethylene oxide (99.7%)	
Hydrogen chloride (99.0%)	Inorganic acid
Methyl chloride (99.5%)	Chlorinated hydrocarbon

Source: ASTM F 1001-86.

B. Permeation Testing

Permeation of a liquid or vapor through protective clothing material, generally an elastomeric rubber or plastic or composite material, involves three steps:

1. The sorption of the chemical at the outside surface of the CPC material
2. The diffusion of the chemical through the CPC material
3. The desorption of the chemical from the inside surface of the CPC

The rate at which chemicals permeate the CPC is of prime importance in selecting appropriate CPC. Also, the time lapse between contact with the chemical and the appearance of the chemical on the inside of the CPC, called *breakthrough time*, should be considered in selecting CPC. A standard test method ASTM F739-85 has been promulgated by ASTM Committee F-23 to test permeation of CPC. The chemical permeation rate depends upon the

1. Diffusion coefficient of the permeating chemical in the CPC material
2. Solubility of the permeating chemical in the CPC material
3. Difference in the chemical concentration between the inside and outside surface of the CPC material
4. Thickness of the CPC material
5. Area of the CPC material in contact with the chemical

Breakthrough time is measured readily by permeation testing, from the initial contact time of the outside surface of CPC with the chemical and the time the chemical is detected on the inside of CPC. Breakthrough time may be the single most important criterion for CPC selection in cases where, for example, a carcinogen is to be handled. However, measurement of breakthrough time depends upon the sensitivity of the analytical test method; hence, caution must be used in interpreting results.

C. Factors Affecting Permeation Assessment

Several factors affect permeation and chemical resistance properties of CPC materials. The following factors should be considered when selecting CPC materials.

1. Temperature

Permeation rates increase, and breakthrough times decrease with increasing temperature. The extent to which barrier properties of CPC will be affected with increasing temperature is chemical/material pair specific.

2. *Material Thickness*

Permeation is inversely proportional to thickness, and breakthrough time is directly proportional to the square of the thickness. Thus, doubling the thickness will quadruple the breakthrough time [10].

3. *Solubility Parameter*

Solubility is measured by the amount of the chemical that can be absorbed by a given amount of CPC material. In general, chemicals having high solubilities will permeate the CPC material rapidly. However, caution must be used in interpreting solubility data, since permeation rate is a function of both solubility and diffusion coefficient. Gases, for example, have low solubilities but high diffusion coefficients and may permeate CPC materials at very high rates compared to liquids.

4. *Multicomponent Liquids*

Selection of appropriate CPC becomes a difficult problem when the chemicals are not known or mixtures of chemicals are encountered. Chemical spills or fire situations involving chemicals are examples of such scenarios for emergency response teams. Mixtures of chemicals are generally more damaging toward CPC materials; also a rapidly permeating component of a mixture may provide a pathway that accelerates the permeation of the other component that would permeate at a slower rate if in pure form [10].

5. *Persistent Permeation*

After a chemical has begun to diffuse into the CPC material, a reservoir is built within the CPC matrix. The reservoir will continue to diffuse even after the chemical on the outside of the surface is removed. This means that even after cleaning the CPC, a chemical may continue to diffuse to the inside of the CPC; hence, there is a hazard in reusing such a CPC item.

6. *Other Factors*

Other factors include the design and construction of chemical protective clothing items and ensembles. Stitched seams, for example, may permit chemical penetration and permeation; hence, they must be sealed with a coating. Defects like pinholes and variations in thickness may undermine protection. Also, the method of manufacturing a CPC item may influence its barrier effectiveness. For example, solvent-dipped gloves produced by a multiple dip process have several layers of the elastomer covering the imperfections of the previous layer; hence, they perform better than single-dip operation latex gloves.

The degree of protection provided by CPC also is a function of duration and type of activity that may require specific physical properties, such as puncture resistance and tear resistance. High temperature and relative humidity cause perspiration, which affects the permeation rate of chemicals through CPC as well as worker fatigue; hence, the period of safe and effective worker activity may be reduced [10].

D. Performance Standards for CPC

The ASTM succeeded in developing two standards for measuring chemical resistance performance of clothing, that is, permeation and penetration resistance. Despite the development of these standards, few products have been tested in accordance with the standard test methods. Also, inconsistent results and information provided in the past have created confusion in the selection of appropriate CPC. The U.S. Environmental Protection Agency (EPA) recommends various types of chemical protective clothing for a particular situation. Potential exposure situations are divided into four categories, each requiring a different level of protection. These include:

1. Severe respiratory, skin, or eye hazard (Level A). For this type of situation, a totally encapsulating chemical protective suit with pressure demand self-contained breathing apparatus (SCBA) or supplied air respiratory with escape SCBA; chemical-resistant gloves and boots are recommended.
2. Severe respiratory hazard and moderate skin hazard (Level B). chemical-resistant coveralls, one- or two-piece splash suit, gloves, and boots; pressure demand SCBA or supplied air respiratory with escape SCBA should be used.
3. Moderate respiratory or skin hazard (Level C). Chemical-resistant coveralls, one- or two-piece splash suit, gloves, and boots; full-face shield and air-purifying, canister-equipped respirator are recommended.
4. No respiratory hazard and mild skin hazard (Level D). Coveralls, safety boots, safety glasses, and hard hats are recommended.

However, the EPA does not define the specific performance of CPC items [11]. There are some other publications [12–14] that provide guidelines for selecting CPC, but these too do not provide information on the performance of ensembles.

In 1984, a series of transportation and industrial accidents involving hazardous chemical exposure to emergency responders led the National Transportation Safety Board to recommend that federal agencies assist in the formation of comprehensive standards to improve the performance of chemical protective clothing. An initial effort began in ASTM for creating comprehensive specifications to document protective suit performance and manufacturer reporting requirements. In 1986, the National Fire Protection Association (NFPA) established a subcommittee on hazardous chemical protective clothing within its Technical Committee on Fire Fighters Protective Equipment. The NFPA subcommittee's work resulted in one draft ASTM standard, ASTM F23.50.03 (Z1416-Z) "Standard Guide for Minimum Documentation and Test Methods of Chemical Protective Clothing," and three NFPA Standards. These three standards took effect in 1990 and are

1. NFPA 1991 Standard on Vapor Protective Suits for Hazardous Chemical Emergencies
2. NFPA 1992 Standard on Liquid Splash Protective Suits for Hazardous Chemical Emergencies
3. NFPA 1993 Standard on Protective Suits for Non Emergency, Non Flammable Hazardous Chemical Operations

In each standard, the suit is defined as being worn in conjunction with a self-contained or air-line breathing apparatus for respiratory protection.

The proposed ASTM Standard F23.50.03 (ASTM-Z 1416Z) has been designed to cover all forms of protective clothing in a general standard with guidelines for manufacturers' minimum documentation of their products. This draft is still in the review process as of April 1993. The obvious benefit to users would be that they can compare products easily on the basis of the same available information. However, providing product documentation consistent with this ASTM standard does not imply an acceptable level of performance for the garment or clothing item. Therefore, the burden of determining acceptable performance is placed on the end user who must be capable of interpreting the data to make an appropriate clothing selection.

III. PROTECTION FROM THERMAL HAZARDS

The primary function of thermal protective clothing is to minimize or eliminate physical harm as a result of fire or exposure to hot surfaces, molten metal splashes, etc. The performance of thermal protective clothing depends on its ability to insulate and to maintain structural integrity when exposed to high heat assault. There are high demands on the materials used under such circumstances. Indeed, different levels of heat exposure are encountered in different industrial work situations. However, no amount of protective clothing can help in flash-over fire situations for more than a few seconds.

A. Characteristics of Textiles for Thermal Protective Clothing

Heat exposure in fire situations may consist primarily of radiation, but convective and conductive heat (if, for example, molten metal or burning parts of structures fall on a garment) also may be encountered. Under any of these conditions, the garments should not ignite; they should remain intact, that is, not shrink, melt, or form brittle chars that may break open and expose the wearer; and they should provide as much insulation against heat as is consistent with not diminishing the wearer's ability to perform.

Fire fighters fighting a room fire have been reported to be exposed to up to 12.5 kW/m^2 and up to 300°C temperatures for a few minutes [15]. Greater exposures can occur in emergencies, such as a falling burning roof [16]. Garment characteristics that protect the wearer from burn injury are many, but the major protective property is thermal resistance, which, in the first approximation, is related to fabric thickness. Also, for a given thickness, the lower the density, the greater the thermal resistance [17]. Moisture has been known to reduce a fabric's heat resistance. Also, heat resistance can be reduced by high temperature, especially if the fibers shrink, melt, or otherwise deteriorate. Furthermore, thermal resistance decreases with curvature, thus requiring more fabric thickness to protect fingers than large body areas [17].

Thermal inertia (density × heat capacity × thermal conductivity) is another factor in reducing heat flux through garments when the exposure is of high intensity and short duration [17]. For most organic fibers (cotton, wool, Kevlar, Nomex, etc.), as opposed to inorganic fibers like glass or asbestos, thermal inertia is proportional to the fabric weight. Since garments retain heat, they should be removed as soon as possible after exposure and the skin cooled, especially if pain is already felt; otherwise, burn injury may result. The time interval between occurrence of pain and a second-degree burn can be very short [18]. Reflective surfaces are very effective in providing heat protection. Thus, surface temperatures of fabrics exposed to radiation are reported to be reduced to about one half in still air by the use of aluminized surfaces with 90% reflectivity and to be reduced considerably more in moving air [17]. However, aluminized surfaces lose much of their effectiveness when dirty [19].

Moisture present in a heat protective garment cools the garment, but it also may reduce the thermal resistance and increase the heat stored in it [17]. If the garment gets hot enough, steam may form on the inside and cause burn injury. In the United States, most fire fighters' turnout coats use a vapor barrier either on the outside or between the outer shell and the inner liner. The barrier prevents moisture and many corrosive liquids from penetrating the garment, but it interferes with the escape of moisture from perspiration and increases the heat stress. Furthermore, tar and gases formed during fabric decomposition transfer heat from the garment to the skin [20,21].

B. Methods of Testing Heat Protective Clothing

1. *Ignition Resistance and Flammability*

Ignition resistance of fabrics used in heat protective clothing can be measured in various ways. One method involves exposing the bottom edge of a vertical specimen to a flame for 12 sec (Method 5903, Flame Resistance of Cloth, Federal Test Method Standard 191). The NFPA Standard 1971, Protective Clothing for Structural Fire Fighting 1981, specifies that, in using this method, the length of the charred material must not exceed 100 mm and flaming must cease within 2 sec after the flame is removed. This test, in one way, is severe because both sides of the specimen are exposed to the flame. But the test does not account for preheating and drying of a garment exposed in a fire situation. Actual fire conditions are simulated more closely in tests in which the specimen is exposed to radiation. The time to ignition, with or without pilot ignition, is reported [22,23].

2. *Limiting Oxygen Index*

The limiting oxygen index (LOI) test is used to characterize flammability of heat protective fabrics [15]. The LOI measures the minimum oxygen concentration (in an oxygen–nitrogen gas mixture) at which materials continue to burn [15,24]. Normally, materials with an LOI > 21 do not burn in air, whereas those with an LOI < 21 do.

The pyrolytic characteristics of selected textile fibers are presented in Table 2 [25].

3. *Heat Protective Properties*

Heat protective properties of garments can be measured, in the first approximation, by the thickness of the material. All methods that measure actual heat protection employ a radiant or flame heat source, a specimen holder, and a heat sensor placed on the side opposite the heat source. One such method is the ASTM Test for Thermal Protective Performance of Materials for Clothing by Open-Flame Method ASTM - D4108-87. In this test, the specimen is tested in a horizontal position. A gas flame impinges on the lower surface of the horizontal specimen. The heat sensor is a copper disk with four thermocouples embedded. When the flame is properly adjusted, the sensor reading 50 mm above the burner rim, without a specimen, corresponds to 84 kW/m^2. The method has been modified to allow a mixture of radiant and convective heat.

Table 2 Pyrolitic Characteristics of Selected Textile Fibers

Fiber	Temperature, °C		Heats of combustion, BTU/kg	Limiting oxygen index
	Decomposition	Ignition		
Wool	230	590 Self-extinguishing	20,833	25.2
Cotton	305	400	16,313	17–20
Asbestos	—	Noncombustible	—	—
Acetate	300	450	16,975	18.4
Rayon	177–240	420	16,313	18.6
Glass	815	Noncombustible	—	—
Acrylic	287	530	2,863	18.2
Aramid		>427	—	—
Modacrylic	235	Self-extinguishing	—	—
Novoloid	[a]			
Nylon	345	532	28,549	20.1
Olefin (propylene)	—	—	—	18.6 —
Polyester	390	560	20,502	20.6
Saran	76	Self-extinguishing	—	—
Vinyon	—	Self-extinguishing	—	—

[a]Decomposes, resists over 2,760°C; does not burn.
Source: Adapted from Ref. 25.

Based on this method, a thermal protective performance (TPP) rating can be calculated for a material or combination of materials. Also, the effect of moisture in clothing and the effect of different radiant/convective heat loads on fabrics can be measured. The multilayered clothing ensemble is exposed to a calibrated radiant/convective thermal flux, and a copper disk containing thermocouples records the heat transmitted through the material. The temperature curve then is compared with a second curve that relates time to second-degree burns. Thus, the TPP rating is a direct measure of the thermal protective capabilities of the clothing worn by a fire fighter [25,26]. Table 3 presents TPP ratings of selected ensembles and single fabrics.

Another apparatus consists of a radiant quartz panel, a vertical specimen holder, and a water-cooled heat flux gage behind the specimen [27–30]. To obtain a specified radiative/convective heat ratio, air can be blown through slots on the periphery of the quartz panel toward the specimen.

Several other methods have been reported in the literature. For example, the military uses a JP-4 burner as a laboratory test [31] and uses Thermoman,

Table 3 TPP Rating for Protective Clothing (100% radiant heat) at 2 cal/cm^2/sec

Ensembles	Total weight, g/m^2	Total thickness, mm	TPP rating
Nomex III shell, neoprene polycotton-vapor barrier, Nomex quilt liner	899	5.6	42.5
Nomex III shell, Gore-Tex vapor barrier, Nomex quilt liner	726	5.8	44.7
Aluminized Nomex I shell	302	0.4	67.7
Aluminized Kevlar shell	346	0.5	78.0

Source: Adapted from Ref. 26.

a mannequin with about 110 thermosensors that is pulled through jet fuel pool fires, for full-scale testing [15]. Another method uses a simulated skin thermosensor that has thermal properties closer to those of skin than copper disks or heat flux gages [32]. The temperature rise during a given time period is used to compare fabrics [16].

C. Materials For Heat Protective Clothing

High temperature-resistant fibers suitable for protective clothing are flame retardant (FR) wool, FR cotton, FR rayon, Nomex, Kevlar, polybenzimidazole (PBI), polyphenylene sulfide (PPS), oxidized acrylic, and carbon. Particularly important to the aerospace program are PBI and PPS. Fiberglass and asbestos are heat and fire resistant, but new high-performance organic fibers are taking over these fibers.

In working environments with high heat radiation, such as the steel industry, forge shops, melting shops, and fire and utility maintenance situations, heat resistance and heat reflectance are desirable properties for protective clothing. To combat radiant heat, aluminum coating is used with woven, felt, or nonwoven materials made of the FR cotton, FR rayon, FR wool, oxidized acrylic, carbon, aramid, and other FR fibers.

Fire fighters' protective clothing, for example, consists of work uniforms constructed of single-layer, light to medium weight FR cotton, modacrylic, aramid, wool, or polyester, whereas firefighters' turnout coats for structural fires consist of an assembly of an outer shell, a vapor barrier, and an inner thermal barrier. The shell usually contains an aramid fiber (Nomex III, Kevlar, or a blend of Nomex with Kevlar, Novoloid, or PBI). The vapor barrier may be composed of neoprene-coated aramid fabric, Gortex-laminated nylon, or neoprene-coated aramid needle punch fabric. The thermal barrier is generally aramid batting, aramid needle bonded fabric, quilt fabric, Nomex/Kevlar felt, or wool felt. One major problem with the turnout coat is its heavy weight (70–80 pounds). It is important that weight be reduced yet a minimum TPP rating of 35 be maintained to enhance mobility and perhaps reduce fatigue.

For the metal industry, Zirpro-treated wool, Caliban-treated cotton, aluminized cotton, and rayon are used for outer coats and garments. Mittens and gloves are usually made of leather, fiberglass, Zetex (a highly texturized fiberglass), etc. In assessing clothing worn in accidents, it has been found that trappings of metal within garments cause injuries [33]. Therefore, it is important to design garments to avoid this possibility.

IV. PROTECTION FROM MECHANICAL HAZARDS

Ultrahigh-strength fibers are increasingly used in a variety of protective clothing applications including ballistic armor and cut-resistant clothing items. Nevertheless, clothing design is important in the prevention of injuries that occur from the clothing item hampering work activity or getting caught in machinery. There are many industrial and farming injuries that are clothing related. The design of clothing is not within the scope of this discussion since it is activity specific. However, general characteristic requirements and polymer materials used for ballistic armor and cut/puncture-resistant materials will be discussed here.

A. Armor Systems

Ballistic armor systems are of two kinds. The first type is ballistic clothing or soft armor that uses the high flexibility, high modulus, tenacious, and energy absorption capacity of high-performance fibers such as Kevlar aramid or Spectra 1000 high molecular weight, extended-chain polyethylene fiber.

The earliest types of armor were generally made of hard and rigid metal, which spread the impact load but were uncomfortable and heavy. With the invention of synthetic fibers, great improvements in armor design have been possible. The first large-scale use of synthetic fibers in armor systems was woven nylon used in combination with steel plates to produce "flak jackets" during World War II [34]. The next improvement came when glass fiber and resin composite armor called Doron replaced the steel plates [34]. Aramid fibers (Kevlar) provided a major improvement in armor system technology because of their balance of physical properties. Aramid armor systems perform well; however, with the invention of high-performance polyethylene (HPPE) fibers (Spectra 900, 1000), it is possible to produce armor systems

Table 4 Selected Properties of Fibers Used as Armor

Property	HPPE Spectra 1000	Aramid HM	S-glass
Density	0.97	1.44	2.49
Elongation, %	2.7	2.5	5.4
Tensile strength, dN/tex	5.04	21.70	20.87
Tensile modulus, dN/tex	2013.66	1030.88	407.91

Source: Adapted from Ref. 35.

that provide protection equivalent to other armor systems but at a lower weight, or alternatively a significantly higher level of protection at the same weight. Table 4 shows selected properties of HPPE, aramid, and glass fibers.

The chemical resistance of the HPPE fibers also is an advantage for military body armor. Chemical resistance is important if the armor is exposed to chemical agents, especially those used in decontamination during chemical or biological warfare [35].

Table 5 shows chemical properties of HPPE and armid fibers toward selected chemicals.

In a typical ballistic vest, the ballistic resistance is obtained by using multiple layers of a woven fabric. The number of layers needed is determined by the level of the threat that must be stopped [35]. The higher the threat, the greater the number of fabric layers required to defeat it. Fabric construction and yarn size also influence the ballistic performance of the armor system. Conventional fabric constructions using high-performance fibers can be penetrated by ballistic projectiles when the yarn in the projectile path is pushed aside without ever fully engaging the projectile. Lower denier (finer) yarns can reduce this tendency. Fabric construction is another factor in the movement of yarns. Plain weave performs better than satin or twill weaves. Nonwoven fabric construction performs even better.

The most common type of hard armor used as protective clothing is helmets, for example, police riot helmets or those used for military purposes. Again, armaid fibers or HPPE are used in composite form to produce good ballistic load absorption. Before the availability of high-performance organic fibers, riot helmets were made from a fiberglass and epoxy composite that provided little ballistic protection. Among the aramid and HPPE fibers, the latter provide superior impact performance at a lighter weight [35].

Table 5 Chemical Properties of Spectra 900 and Kevlar

	Percent strength retention after 6-months immersion	
Chemical	Spectra 900	Aramid
10% detergent solution	100	100
Perchlorethylene (dry cleaning solvent)	100	75
5*M* sodium hydroxide	100	42
Ammonium hydroxide (29%)	100	70
Hypophosphite solution (10%)	100	79
Clorox	91	0

Source: Adapted from Ref. 35.

The type and amount of the high hardness material required is heavily dependent on the type of threat that must be defeated. Use of high-performance organic fibers in hard panels can improve the performance of soft body armor by protecting vital areas of the body such as the chest. Hard body armor also has sporting applications such as protective head gear.

B. Cut/Slash/Puncture-Resistant Protective Clothing

For employers, protecting their workers from injury has become a primary concern since the passage of the Health and Safety at Work act of 1974 [36]. This act made employers responsible for providing adequate protective clothing and equipment where necessary. One of the most susceptible parts of the body to injury is the hands. Even by protecting the hands, by wearing gloves and installing guards, hand and finger injuries account for over 25% of all industrial accidents [36]. Clearly, there is a need for improved types of hand protection.

1. Cut-Resistant Gloves

Almost all gloves can be considered to provide some degree of cut resistance because the glove is an additional material that must be cut through before the cutting edge reaches the hand. Cotton, leather, and terry cloth gloves all provide some protection from cuts. However, products that have been designed for use in a work environment where there is a high potential for laceration must offer a very high level of cut resistance in order to provide maximum protection. Cut-resistant gloves are constructed from glass fiber,

Kevlar, HPPE, carbon fibers, nylon, and polyester. In many cases, composites rather than single fibers are used to obtain a high level of cut resistance. Nevertheless, HPPE and aramid fibers perform very well for such applications because of high tenacity and modulus, good abrasion resistance, and lubricating properties that permit sharp edges to slide away. The HPPE properties permit construction of seamless gloves by knitting technology.

Besides strength and abrasion-resistance properties, HPPE can be sanitized readily. For food processing, the ability to disinfect gloves is especially important. The chemical properties of HPPE and aramid fibers allow the use of a variety of disinfectants including bleach in refurbishing the gloves for reuse. Also, the low-moisture regain of these fibers, especially that of HPPE reduces the probability of bacterial growth.

2. *Protective Sweaters*

High-performance fibers are used to produce knit sweaters that have excellent abrasion- and cut-resistant properties for motorcycle riders and may find applications in law enforcement or police work. The sweater yarns are composite type, that is, HPPE or Kevlar wrapped with nylon or polyester for ease of dyeing. The sweater protects the wearer from scrapes and abrasions and against knife slashes, but not against punctures such as knife stabs because of the open nature of the knitted construction [35].

3. *Medical Glove Liner*

The need has never been more acute for better hand protection for health care professionals. In the past, the main reason for using sterile gloves and clothing was to protect the patient from the doctor. Today, this situation is, in many cases, reversed. Because of the Acquired Immune Deficiency Syndrome (AIDS) virus, the need to protect health care workers from the patient has become a major concern. Another less well publicized but far more prevalent threat to health care workers is the danger posed by other infectious diseases such a the hepatitis B virus. The Occupational Safety and Health Administration estimates that approximately 300 health care workers die annually from hepatitis B infections [37]. Many of these workers contract infectious diseases through cuts on their hands and fingers. Another route of infection is through accidental needle sticks [37].

The first extra precaution taken by health care professionals against accidentally infecting themselves was to wear multiple pairs of latex rubber gloves [38]. The ideal situation would be to provide a glove that would isolate the hands from the environment and at the same time protect the hands from accidental nicks and cuts. The HPPE fiber glove liner provides a high level of cut protection (for surgical staff) while affording enough tactile sensitivity. Unfortunately, the HPPE fiber glove liner, because of its knitted construction, does not provide protection from accidental needle sticks.

The glove liner can be reused because it is separate from the latex glove. The liner can be sterilized by a variety of methods such as ethylene oxide, radiation, steam autoclave, or bleach. However HPPE cannot be sterilized at higher than 121°C because of the fiber's low melting temperature.

4. Chain Saw Cut Protection

Early chain saw cut protective chaps were made with over 22 layers of ballistic nylon. Later chaps were produced with multiple layers of aramid material (Kevlar), and currently HPPE chaps are made from two layers of nonwoven fabric. The nonwoven fabric construction is capable of protecting the user from injury from chain saw cuts at chain speeds in excess of 3200 ft/min [35].

V. PROTECTION FROM BIOLOGICAL HAZARDS

Due to the federal government warnings and currently mandated protection of employees from contagious diseases, protective clothing is center stage in the health care industry. Protective clothing is essential for hospital support staff in operating rooms, isolation areas, emergency rooms, and intensive care units. Other work situations requiring protective clothing against biological hazards are law enforcement, ambulance, blood laboratory, and even mortuary workers.

The basic functions of apparel worn in the operating room are to help preserve a clean environment, reduce the number of airborne microorganisms, help preserve sterile or aseptic operating room techniques, protect surgical personnel from pathogens originating from the patient, and protect the patient from bacterial transmissions from surgical personnel.

Since untreated cotton operating garments have been found to contribute to higher particle counts in the operating room and not provide an effective sterile barrier between the surgical personnel and the patient, they are rapidly disappearing. One of the major attributes of the cotton operating gown, however, is that it is very comfortable to wear.

One of the techniques used to improve the comfort of operating gowns is to use an impermeable material (barrier) for the front half of the gown and to use a more breathable fabric for the back. The AIDS epidemic has resulted in a need for more effective gowns, lab coats, coveralls, aprons, and other medical garments for the protection of hospital and medical laboratory personnel.

A. Physical Methods of Imparting Protection

Nonwoven fabric manufacturers have developed a line of clothing products called *Control Cover*. The products are made of three layered composite fabrics (spunbonded/melt blown/spunbonded). The manufacturer has devised a series of tests to measure blood strikethrough for gowns. Control Cover clothing shows only a 1.4% strikethrough compared to ranges from 68.6 to 82.9% for other gowns.

Spunlaced fabrics based on polyester/pulp combinations and composites of spunbonded polypropylene/melt blown polypropylene/spunbonded polypropylene are two of the major fabrics used in disposable surgical gown and drape applications.

Baxter Health Care Products is using DuPont's Sontara polyester/pulp fabrics in its surgical gowns and drapes. The Burgikos Division of Johnson & Johnson is reported to be changing some of its surgical gown and drape lines from the Sontara fabric to a polypropylene composite fabric.

Wet-laid nonwoven products have found a large number of applications in the hospital/medical field. Dexter nonwovens has its Stasis line of polyethylene-coated fabrics. Included in this line are splash mask coverstock, polyethylene-coated emergency drape/sheets, polycoated yellow- and blue-colored fabrics for shoe covers, lab coats, emergency drapes, cover gowns, and face masks. Another advantage of coated nonwovens is that they are lint free, since lint from clothing may act as carrier of microbes, bacteria, and viruses.

The older cotton gauze mask with its 50% or less efficiency has been superseded by disposable nonwoven masks that provide efficiencies as high as 98%. Melt blown microfiber filter media is displacing much of the microglass media that had been used for surgical face masks.

B. Chemical Methods of Imparting Protection

Antimicrobial chemicals can be coated to all types of woven or nonwoven materials in a polymeric form so that they are durable to repeated laundering or can be microencapsulated within the fiber with controlled release properties. Examples of such protective clothing materials are discussed in detail by Vigo in Chapter 9 of this volume.

C. Decontamination and Sterilization

Decontamination of reusable clothing can be achieved by laundering in hot water with detergent and adding disinfectants such as bleach. A variety of other disinfectants may be suitable for removing bacteria and other microorganisms. For example, peroxides, peracids, perborates, or halogenated guadinium salts can be used.

There are three basic methods of sterilizing products for use in hospital/medical applications: steam sterilization, ethylene oxide sterilization, and gamma ray sterilization. The third method, gamma ray sterilization is gaining in popularity.

The gamma ray sterilization process has many advantages over the other systems in terms of thoroughness of treatment. There are, however, some problems with the effect of gamma rays on some materials. Cellulosic materials experience degradation and produce some odors. High-density polyethylene loses some strength. Polypropylene, nylon, and polyvinyl chloride undergo some degradation, and polypropylene also discolors. Polyester fibers and acrylic binders are unaffected by the gamma rays.

VI. CONCLUSIONS

The majority of workers with a potential for exposure to health and safety hazards are protected by using some form of personal protective clothing (PPC) and equipment. There is a variety of PPC available for specific work situations. However, selection of the most appropriate form of protection is very complex, as safety, mobility and dexterity, comfort, and cost must be balanced.

General awareness regarding health and safety of workers and government regulations and standards have prompted much research and development in the areas of polymer materials, design of PPC, test methods and performance evaluation techniques, and international cooperation. However, much work needs to be done so that manufacturers of PPC may use uniform labeling system for protective clothing items with safety data and recommendation for use in specific work environments.

REFERENCES

1. D. N. Eiser, Problems in Personal Protective Equipment Selection, *Performance of Protective Clothing*, ASTM STP 989 (S. Z. Mansdorf, R. Sager, and A. P. Nielsen, eds.), American Society for Testing and Materials, Philadelphia, 1988, pp. 341–346.
2. A. M. Stoll and M. A. Chianta, Method and rating system for evaluation of thermal protection, *Aerospace Medicine 40*:1232 (1969).
3. W. P. Behnke, Thermal protective performance test for clothing, *Fire Technology 13*:6 (1977).
4. B. E. Pintauro and R. L. Barker, A summary of research on heat resistant fabrics

for protective clothing, *American Industrial Hygiene Association Journal 44*:123 (1983).
5. W. F. Baitinger, Product engineering of safety apparel fabrics: Insulation characteristics of fire retardant cottons, *Textile Research Journal 49*:221 (1979).
6. Flame resistance of cloth; Vertical, Method 5903, U.S. Federal Test Method Standard No. 191, Textile Test Methods (1979).
7. Occupational Safety and Health Administration, *Fed. Register 56* (235):64004 (Dec 6, 1991).
8. M. Hougaard, The need for international cooperation regarding approvals, *Performance of Protective Clothing*, ASTM STP 989 (S. Z. Mansdorf, R. Sager, and A. P. Nielsen, eds.), American Society for Testing and Materials, Philadelphia, 1988, pp. 7–14.
9. J. O. Stull, R. A. Jamke, and M. G. Steckel, Evaluating a new material for use in totally encapsulating chemical protective suits, *Performance of Protective Clothing*, ASTM STP 989 (S. Z. Mansdorf, R. Sager, and A. P. Nielsen, eds.), American Society for Testing and Materials, Philadelphia, 1988, pp. 847–861.
10. A. D. Schwope, P. P. Costas, J. O. Jackson, and D. J. Weitzman, *Guidelines for the Selection of Chemical Protective Clothing*, Vol. I, Arthur D. Little, Cambridge, Mass., 1983.
11. J. O. Stull, Performance standards for improving chemical protective suits, *Chemical Protective Clothing Performance in Chemical Emergency Response*, ASTM STP 1037 (J. L. Perkins and J. O. Stull, eds.), *American Society for Testing and Materials*, Philadelphia, 1989, pp. 245–263.
12. *1987 Emergency Response Guidebook, DOT P 5800.4*, U.S. Department of Transportation, Research and Special Programs Administration, Office of Hazardous Materials Transportation (1987).
13. A. D. Schwope et al., *Guidelines for the Selection of Chemical Protective Clothing*, 3rd Ed., ACGIH, Cincinnati, Ohio, 1987.
14. K. Forsberg and L. H. Keith, *Chemical Protective Clothing Performance Index*, Wiley-Interscience, New York, 1989.
15. N. J. Abbott and S. Schulman, Protection from fire: Non-flammable fabrics and coatings, *Journal of Coated Fabrics 6*:48 (1976).
16. J. F. Krasny, Some characteristics of fabrics for heat protective garments, *Performance of Protective Clothing*, ASTM STP 900 (R. L. Barker and G. C. Coletta, eds.), American Society for Testing and Materials, Philadelphia, 1986, pp. 463–474.
17. D. L. Simms and P. L., Hinkley, Protective clothing against flame and heat, *Fire Research Special Report No. 3*, Joint Fire Research Organization, Department of Scientific and Industrial Research and Fire Officers Committee, Her Majesty's Stationary Office, London England, 1960.
18. A. M. Stoll and M. A. Chianta, Heat transfer through fabrics as related to thermal injuries, *New York Academy of Science, Transaction 33*:649 (1971).
19. J. Quintiere, Radiative characteristics of fire fighter's coat fabrics, *Fire Technology 20*:153 (1974).

20. J. M. Davies, B. McQue, and T. B. Hoover, Heat transferred by decomposition products from cotton fabrics exposed to intense thermal radiation, *Textile Research Journal 35*:757 (1965).
21. W. F. Baitinger, Product engineering of safety apparel fabrics: Insulation characteristics of fire-retardant cotton, *Textile Research Journal 49*:221 (1979).
22. M. M. Schoppe, J. M. Welsford, and N. J. Abbott, Resistance of Navy shipboard work clothing materials to extreme heat, *Technical Report No. 148*, Navy Clothing and Textile Research Facility, Natick, Mass., 1982.
23. J. F. Krasny, P. J. Allen, and A. Maldonado, Burn injury potential of navy shipboard work clothing, *Technical Report No. 146*, Navy Clothing and Textile Research Facility, Natick, Mass., 1983.
24. E. P. Brewster and R. L. Barker, A summary of research on heat resistant fabrics for protective clothing, *American Industrial Hygiene Association Journal 44*:123 (1983).
25. J. Yeager, *Textiles For Residential and Commercial Interiors*, Harper & Row, New York, 1988.
26. J. H. Veghte, Functional integration of fire fighters' protective clothing, *Performance of Protective clothing*, ASTM STP900 (R. L. Barker and G. C. Colleta, eds.), American society of Testing and Materials, Philadelphia, 1986.
27. E. Braun, D. Cobb, V. B. Cobble, J. F. Krasny, and R. D. Peacock, Measurement of the protective value of apparel fabrics in a fire environment, *Journal of Consumer Product Flammability 7*:15 (1980).
28. R. M. Perkins, Insulative values of single-layer fabrics for thermal protective clothing, *Textile Research Journal 49*:202 (1979).
29. R. M. Perkins, J. F. Krasny, and E. Braun, Insulative values of double layers of fabrics exposed to radiative heat, *Proceedings, 13th Annual Meeting, Information Council on Fabric Flammability*, Atlanta, Ga., 1979, pp. 88–96.
30. R. M. Perkins, J. F. Krasny, E. Braun, and R. D. Peacock, An evaluation of fabrics for thermal protective clothing, *Proceedings, 12th Annual Meeting, Information Council on Fabric Flammability*, New York, 1978, pp. 212–230.
31. J. H. Ross, Thermal conductivity of fabrics as related to skin burn damage, *Journal of Applied Polymer Science: Applied Polymer Symposium 31*:293 (1977).
32. R. C. Maggio, A molded skin simulant material with thermal and optical constants approximating those of human skin, NS 081-001, Naval Material Laboratory, Naval Shipyard, New York, 1956.
33. Fire report on clothing to protect against molten metal splash hazard in foundries, Her Majesty's Stationary Office, Health and Safety Commission, London, 1985.
34. T. A. Abbott, Ballistic protective clothing, *Shirley Publication S 45*:41 (1988).
35. K. M. Kirkland, T. Y. Tam, and G. C. Weedon, New third-generation protective clothing from high-performance polyethylene fiber, *High Tech Fibrous Materials* (T. L. Vigo and A. F. Turbak, eds.), American Chemical Society, Washington, D.C., 1991.

36. D. C. Bennett Hand protection, *Health and Safety at Work*, 32 (May 1981).
37. New products to prevent accidental needle sticks, *Sunday New York Times* (March 27, 1988).
38. P. Truell, Demand for rubber gloves sky rockets, *The Wall Street Journal*, p. 6 (Thursday, June 9, 1988).

2
Design of Protective Clothing

ELIZABETH P. EASTER University of Kentucky, Lexington, Kentucky

I. INTRODUCTION

Clothing is a basic human need that traditionally is viewed as a means of satisfying the aesthetic needs of fashion, but today the need for fashion has been combined with a critical need for function. Today we realize that clothing can function as an important component of human safety. A large number of clothing items function as a means of providing protection for the human body. The relationship of clothing in providing protection from occupational hazards in the users' environment has created a whole new approach to looking at the role of clothing in meeting human needs. According to Watkins [1], clothing is our most intimate environment and is being used to create its own environment, but what makes clothing unique is that it is an environment that is carried everywhere with an individual (i.e., clothing is a portable environment). Clothing designed to provide occupational safety must take into account the effect clothing has on its wearer and the interface between the clothing users and their environment.

One approach to clothing design that has emerged as researchers recognize the need for function as well as fashion is functional clothing design. Functional clothing design is (1) a process (research process) and (2) a product. When an article of clothing is the end product of the functional clothing design process, even the most creative solution is examined from the perspective of the actual world of physical and behavioral resources and the external environment. In the functional design process, the attractive visual effects are worthless if a garment is uncomfortable or does not perform its function. Functional design deals with how something works, how it performs (i.e., how it functions). Using the functional design process, a garment designed for occupational safety will be well designed in three aspects: (1) function, (2) structure, and (3) aesthetics.

In the functional design process, a designer determines requirements for what the garment must or must not do functionally, for example, a pocket that will hold a farmers' keys or a glove that will enable the medical assistant to pick up a surgical instrument quickly. When the functional requirements have been identified, a prototype garment is developed. A prototype is a plan for the structural design of a garment. The structural design of garment determines how it is constructed or how it is put together to fulfill its function. It determines the structural lines, shapes, and parts, in addition to how they will relate to each other, how the garment will fit, and where and how it will open and close. The structural design must not only agree with the garment's function but must also conform with the structure of the human figure. The most successful designs are often those that meet their functional criteria and purpose with the simplest form—that is, form follows function.

Aesthetic design is for appearance only. In the functional design process, it is critical that the aesthetic design not affect the fit or performance of the garment. It must agree with both the functional and structural designs. This means that functional designs or each aspect of a design such as a zipper may also be decorative, but if the design deals exclusively with human protection and safety, the function of a zipper must be ease of donning and doffing the garment with appropriate sealed seams or cover flaps. On the other hand, if the aesthetic needs of the end user have not been met, the garment will not satisfy the needs of the end user. For example, if the protective garment designed to meet the needs of the farmer provides protection from toxic chemicals but is not aesthetically acceptable to the farmer, the garment will not be used except in extenuating circumstance (e.g., required by law for commercial applicators).

II. FUNCTIONAL DESIGN PROCESS

Functional clothing design involves a process that takes the designer step by step from the initial request for a design through the evaluation stage of the final design. The process of functional clothing design combines the theories of physical and social sciences with basic clothing design. It encompasses a holistic or systems approach to designing that had its origin rooted in the ergonomics research conducted through organizations such as the National

Bureau of Standards, the U.S. Army Research Institute, and the National Aeronautics and Space Administration.

The functional design process attempts to externalize the creative thinking of traditional fashion designers by strategizing the design process. A major advantage of bringing design thinking into the open is that other people, such as users, can see what is going on, contribute information, and provide insight to solutions that may be outside the designer's knowledge and experience. The end product of the functional design approach is not only meant to meet specialized clothing needs of the user such as providing a barrier to toxic chemical but also to look at the user's environments including the near as well as the external environments, for example, the heat stress of the user and the climatic conditions under which the user will be working. This approach attempts to accommodate these environments. The solution to the design problem incorporates the knowledge of fashion and human needs into functional designing. Overall, functional clothing is designed to meet the physical, social, psychological, and aesthetic needs of the potential users.

A. Strategy of the Functional Design Process

Table 1 outlines the steps in the functional design process. The process was adapted to apparel design by Jacky DeJonge [1], based on a strategy developed by J. Christopher Jones [2].

1. Request for Design

The functional design process begins with a request for clothing to meet a specific need (step 1). The request may come from the ultimate end user or from a company or firm on the behalf of the end user. For example, the request for a garment for pesticide protection may come from the farmer, a pesticide company, or a commercial applicator firm. In general, the request identifies the problem in very broad or general terms, such as, clothing for a pesticide applicator or a hospital worker in a clean room environment. The request may provide little insight into the problem and usually gives the designer very few specifics that may be used in solving the problem.

2. Exploring the Design Situation

It is at step 2 that the request for the design or design problem is thoroughly explored. The purpose of this step is to explore the design situation and identify as many different directions for investigation as possible. One of the most useful mechanisms to explore the design situation is for designers to observe their clients in action, that is, observe the end users in the environ-

Table 1 Strategy of the Functional Design Process

Step 1.	Request for Design
Step 2.	Exploring the Design Situation
	a. State General Objective
	b. Brainstorming
	c. User Interview and Observation
	d. Literature Search
Step 3.	Analyzing the Critical Design Factors
	a. Observation Analysis
	b. Market Analysis
	c. Literature Search
	d. Identification of Critical Factors
Step 4.	Describing the Specifications
	a. Activity Assessment
	b. Movement Assessment
	c. Impact Assessment
	d. Thermal Assessment
	e. Social–Psychological Assessment
Step 5.	Establishing Design Criteria
	a. Charting
	b. Ranking
	c. Prioritizing
Step 6.	Developing the Prototype
	a. Materials Testing
	b. Technique Evaluation
	c. Creative Integration
	d. Solutions Weighed Against Criteria
Step 7.	Evaluating the Prototype
	a. Specification Testing
	b. User Satisfaction

ment in which the problem was perceived. For example, if a designer observed the design situation of a farmer applying pesticides in a vinyl coated rain suit, the designer could see that the garment would protect the user from chemical exposure but would be uncomfortable in hot weather. Visual observation will enable the designer to have first-hand knowledge of the design situation, the end users and their environments. Another advantage of visual observation is that it enables the designer to look for visual inconsistencies or obvious faults in existing designs.

A second technique used in exploring the design is personal interviews. The designer interviews the end users. Obtaining input from the ultimate end users at the beginning of the design process can enable the designer to avoid pitfalls in the design solution. For example, if the end users (i.e., farmers), felt strongly that a protective garment should have a removable head covering (i.e. hat), then a hooded garment would not meet their needs. Other aspects of the exploration of the design situation may include brainstorming with others or a searching through the literature to find out more about the problem area.

It is important in the design exploration stage that the designer looks beyond the obvious request a client makes when seeking a design solution toward a more comprehensive picture of the clients in their environments. The scope of the solutions to the problem should include all aspects of the constructed, the behavioral, and the natural environments. For example, in designing clothing for pesticide applicators, a thorough observation of the constructed environment brings consideration of the method of pesticide application and the equipment variations to create the human–machine interface. The natural environment focuses attention on the weather conditions during the time of year when pesticides must be applied and creates such critical factor as thermal considerations (Table 2). The behavioral environment introduces the human factors (Table 3), all of which must be important considerations in establishing design criteria.

The users and their environments are examined at this stage of exploring the design situation. The functional designer may choose any or all the strategies in step 2 for completing the exploration of the design situation.

3. Analyzing the Critical Design Factors

After the exploration of the design situation is complete, the designer must thoroughly explore the design problem (step 3). This step involves isolating

Table 2 Environmental Factors

Weather
Temperature
Relative humidity
Wind speed

Table 3 Human Factors

Values	Customs
Attitudes	Preferences
Beliefs	Opinions

the critical factors that comprise the problem. The techniques used for this stage of the design process are taken from basic research methods, including visual observation, literature search, and market analysis. For example, a study conducted by De Jonge et al. [3] used a questionnaire that included an illustration of protective apparel products, with descriptions and fabric samples for each. The purpose of the study was to evaluate the effects of perceived product attributes of functionally designed protective apparel and risk-related factors influencing the adoption/purchase decision. Another approach may be direct observation of an individual in the workplace. For example, observing the pesticide applicator spraying citrus tress in Florida during the month of July when the temperature reached 100°F by noon, the relative humidity was 90% or greater, and the winds were 0–5 mph showed the designer that not only is protection from toxic chemicals important but the thermal comfort of the applicator is also significant.

During this step of the research process the literature search is focused on identifying what has been reported to help in solving the problem for a requested design and isolating critical factors. The number of critical factors and the significance of each will vary with the problem. Upon completion of this stage in the process, critical areas have been identified for further assessment. And, as the results of these strategies converge, the problem is defined.

4. Describing the Specifications

When the problem has been defined, the previously identified critical factors are assessed relevant to the specific problem in order to arrive at specifications for designing (step 4). During step 4, data are gathered for the critical factor areas of the design problem. There are five areas of assessment for identification of the major critical factors: activity, movement, impact, thermal, and social–psychological. All are defined as assessments because they require an in-depth investigation to examine the design problem fully. The number and depth of inquiry of each assessment will vary with the problem being studied.

(a) Activity Assessment. All design problems require some type of activity assessment. This assessment involves a more in-depth observation than was done to identify the critical factors. It is recommended that during the activity assessment the observing of the activity be as it is actually being performed in its natural setting. Also, if possible, observations during this stage should be verified by multiple observers.

There are a number of ways to approach the activity assessment. One is to observe and analyze the activity and then compile a list. To analyze an activity, the complete activity should be observed first, and then its components as well as the frequency with which they occur should be noted. An example of the task analysis approach would be to begin by (1) listing all the tasks included in the job, (2) organizing and arranging the tasks, (3) developing a recording scheme, (4) recording the duration of each task, and (5) ranking their significance. The significance of ranking is where the subjective analysis comes in; hence, this rating may be accomplished by comparing the ratings of a panel of judges. Alternately, the designer may ask the end users how they would rate a specific task. An illustration may be the pesticide worker, in which case the entire operation may be observed and then the activity of mixing or loading may be observed specifically. The specific tasks such as opening and closing a pesticide container may be recorded. Later this type of information will be executed into the design solution (e.g., gloves that provide the user with the dexterity to open and close the container).

An alternative to direct field observations is filming the activity. This method is widely used in the sports field, not just to observe players techniques but also to design clothing for the activity. If conditions do not permit direct observation, a "simulation" of the activity may be enacted and observed. This method is widely used in the space suit-designing process.

Analysis of the information gathered through the activity assessment will result in specifications concerning how the garment can be designed to eliminate interferences with the task involved and where possible to assist in the task. For example, when analyzing the activity of applying pesticides, the user of a suit (farmer) will stop and take a break, maybe smoke a cigarette. How can the garment design ensure that this activity may be carried out without contamination of the inside of the suit and/or the cigarette itself? Resulting specifications for the design may include an outside pocket with appropriate closure to hold the cigarette and lighter or a more functional closure enabling the user to don and doff the garment without contaminating the undergarments or skin.

(b) Movement Assessment. Providing mobility in clothing is an essential component of the functional design process. The success of a protective garment depends a great deal on how easily the user can move about while wearing it. Mobility in clothing is related to both comfort and function. If a client can move in a garment without creating stress or strain, the individual, in general, will be more comfortable. If clothing can move with the individual and the user doesn't have to work against the garment, the individual can perform the task more effectively.

In the past, mobility in clothing was reduced as the level of protection increased. An illustration of an ancient protective garment with limited mobility would be the knight's suit of armor. This lack of mobility was also apparent in the earlier designs of the space suit. Today by using barrier fabrics and advancements in construction techniques, a protective garment can provide both mobility and protection to a space suit or a chemical protective garment.

The first step in conducting a movement assessment is to identify the aspects of movement that are most critical to a user's activities. All types of movement (flexion, extension, lateral, etc.) should be measured. A movement assessment is also conducted to determine whether clothing interferes with the user's performance of a task. Information may be gathered by visual observation of the stress and strain on a garment during movement or by inspecting used garments to determine areas of wear or stress caused by either fabric and/or seam stretch. In addition, a movement analysis may be conducted by filming or videotaping an activity. The film or videotape could be stopped to enable the designer to record accurately the specifics of movement. The specifications resulting from the movement assessment are usually of a very specific nature and are generally met in the execution of the design through both fabric and design features.

(c) Impact Assessment. The term *protective clothing* implies a potentially hazardous working environment. To design protective clothing for occupational safety, the designer must first understand the impact of the hazard, the potential user, and the environment.

The first step in designing for impact protection is to determine what type and level of protection are needed. The role of clothing in providing protection can be determined only if the designer is fully aware of the potential hazard. Hazards may be divided into impact by force and impact by intrusion [1]. Impact by force may be measured by using sensors on simulated body parts to measure blows as in the case of contact sports, or observation methods may also be used to determine the area of the body where blows are most frequently received and a subjective measure of their severity. Force, defined as intrusion of substances on the body, may also be analyzed for the areas of contact and the degree of penetration. For example, an absorbent fabric that changes color upon contact with water allows identification of high-impact areas for clothing worn in the rain.

The functional clothing designer may use direct observation and measurement techniques to identify the specifics for impact protection. That is, the designer may go to the worksite and measure the extent of impact. An example was illustrated by DeJonge et al. [4] in which the overall deposition of pesticides during pesticide application processes was measured by the use of colormetrics (Figure 1). Fenske [5] has taken this concept one step further to illustrate actual human exposure by using fluorescent tracers to demonstrate that substantial pesticide penetration occurred even with the use of protective clothing.

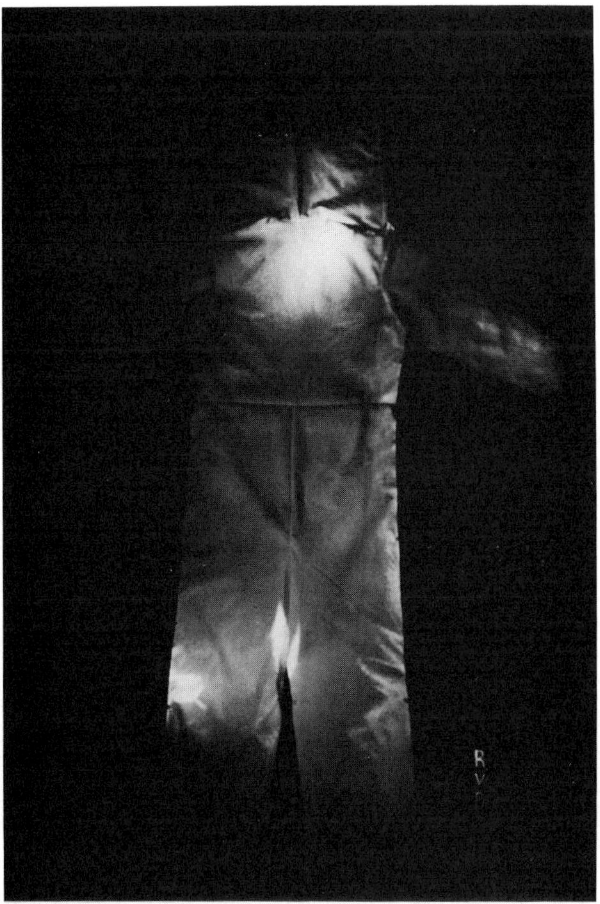

Figure 1 Colormetrics showing pesticide deposition pattern.

Another application of evaluation of impact are field studies [6,7] in which actual pesticide penetration through garments is being measured by the use of absorbent pads worn underneath the clothing. Table 4 identifies some of the strategic locations for the placement of pads in a field exposure study.

A key to impact assessment is to identify the section of the body exposed to the impacting particles. If only one segment of the body will be impacted, then the garment needs to be designed to protect only that portion of the body. For example, x-ray technicians who are exposed to x-rays from the waist up to

Table 4 Recommended Location of Pads

Body region represented	U.S. EPA[a] (1987)	NACA[b] (1986)	WHO[c] (1982)
Head	Shoulder, back, and chest	Front and back	Head
Face	Chest	—	—
Back of neck	Back	—	—
Front of neck	Chest	—	—
Chest/stomach	Chest	Chest	Sternal
Back	Back	Back	Back
Upper arm	Shoulder and forearm	Upper arm	—
Forearm	Forearm	Lower arm	Midway between elbow and wrist
Thigh	Thigh	Upper leg	Thigh
Lower leg	Shin	Lower leg	Lower leg
Feet	—	—	—

[a]Ref. 11.
[b]Ref. 12.
[c]Ref. 13.

the neck will be much more able to do their jobs if the design incorporates lead barrier or whatever mechanism is to be used in the area from the waist to the neck.

If impact protection was the only concern of designers, they would have no problem creating protective garments from the variety of textiles in today's market [1]. But the end users must be able to function under the protection of these garments, which means that the factors identified in the movement assessment are crucial, as are the factors that will be identified in the thermal assessment.

(d) Thermal Assessment. There are two components of thermal assessment: physical and psychological. In physical thermal assessment, thermal factors may be assessed through the use of body sensors worn underneath clothing during the real and/or simulation of an activity. For example, the "environmental chamber" testing of the prototype designs by Branson et al. [8] used the Kansas State University Environment Chamber. This simulated environment enabled the comparative analysis between fabric and garment design variables to be made without elaborate equipment and calculations used in field testing.

Subjective thermal assessment is a determination of the psychological aspects of thermal comfort (i.e., it is based on how comfortable the individual feels). The sensation of comfort and physiological temperature are not necessarily identical and should be evaluated separately. One method of evaluating the psychological component is by direct interview to determine how comfortable individuals feel and how they prefer to feel while engaged in a specified activity.

The results of the psychological assessment may not always agree with the results of the physiological assessment. For example, Nigg et al. [9] reported that during field testing subjects felt that they were not uncomfortable even though their monitors (thermistors) showed that their body temperature had reached a level that required that they be removed from the study.

(e) Social–Psychological Assessment. Of equal importance with the physical assessments are the social–psychological assessments. Research conducted on clothing values and preferences and clothing and social acceptance provides the methodological approach for this area of investigation. The potential users of the design are surveyed to provide insight regarding preferences and values that will affect the ultimate acceptance of the garment to be designed. The specific methods include questionnaires, surveys, and interviews to determine how the users feel and what they think as well as observations of real actions.

The social-psychology of clothing is a subjective area, and, therefore, it is very hard to make generalizations in this area. The researcher has to be very careful in conducting social–psychological research not to mislead the end users.

After the assessments are complete, the data collected from the assessments are then developed into specifications (step 4). Specifications are facts and are sometimes called *user requirements*. With the information gathered in terms of describing specification, it is now time to establish the specifications that will be used in designing—or the design criteria (step 5). The design criteria will enable the researcher to decide how an acceptable design is to be recognized.

5. Establishing Design Criteria

The first stage in establishing the design criteria involves listing or charting the specifications derived in the previous stage. At the onset, the list or chart should include all the specifications. The specifications are now identified as criteria for the design. The second stage in establishing the design criteria is to rank the specifications or determine their weight in order to set a priority for their inclusion into the design solution. Since all the specifications may not be included in the final design, it is important to construct a chart of specifications that are compatible and to use rankings to prioritize and identify the final set of design criteria.

The design criteria are best illustrated using visual representation or pattern language. The purpose is to translate the written specifications into a picture (i.e., take the specifications after chart development, organization, and prioritization and attempt to visualize and represent). Pattern language takes an idea and illustrates a specification (i.e., pattern language is a pictorial representation of the specification). For example, the mixer/loader of pesticide chemicals requires hand protection.

6. *Developing the Prototype*

With a set of design criteria established, this information becomes a guide for the designer in step 6, which is developing the prototype. The prototype development stage includes: materials testing; technique evaluation; brainstorming; creative integration; and solutions weighed against criteria. This stage involves the testing of materials, which may include textile testing, to ensure that the materials selected will meet the criteria established as illustrated in Table 5. For the clothing designed for pesticide protective clothing, a great deal of research has been conducted in this area of the design process. Researchers who evaluate the barrier properties of fabrics provide an example of materials testing for the pesticide user.

Evaluation of construction techniques also may be necessary to see if the seams selected will hold up to the stress specifications and hazardous material penetration. Dimit et al. [10] examined the function of seams and zippers as barriers to pesticide penetration. Table 6 shows the results of the materials testing.

Thus, it is the creative integration of the criteria that leads to possible solutions. These solutions are then evaluated against the prioritized list of criteria to determine what will be incorporated into the final design. The

Table 5 Representative Fabric Characteristics

Fabric	Thickness, mm	Weight, g/m^2	Air permeability, m^3/sec/m^2	Spray rating
Denim*	0.66	276.3	0.13	0
Tyvek	0.10	45.6	0.009	90
Sontara	0.32	72.7	0.298	90
SMS	0.29	62.4	0.056	70

Thickness—ASTM D177-64; weight—ASTM D3776-79; air permeability—ASTM D3737-75; spray rating—AATCC 22-1980; yarn count—ASTM 3775-79; *yarn count—(Denim 71W, 49F).
Source: Adapted from Ref. 10.

Table 6 Percentage of Zippers That Allowed Pesticide Penetration

Zipper description	Percentage allowing penetration[a]
Centered	41
Lapped traditional	8
Lap experimental	18
Open nylon	24
Open metal	16

[a]Percentages are based on 16 replications.
Source: Adapted from Ref. 10.

garment prototype of the design is now ready for construction. It should include the specifications for functional, structural, and aesthetic needs.

B. Evaluation of the Prototype

The important final step 6 is evaluation of the prototype. Evaluation becomes easy with specifications as a guide. To assess whether the garment meets the criteria established, both objective and subjective evaluations should be included (i.e., specification testing and user satisfaction). The researcher must evaluate how the users feel about the solution by the use of a subjective evaluation to complement the objective evaluation.

The evaluation of the prototype may include fitting the garment to a model. This would include testing for fit, design, and comfort. For example, if the design solution was a one size fits all as is the case in many chemical protective suits, the evaluation of the prototype should include a fitting of all the possible size ranges that may be using the final garment.

The evaluation of the prototype may also include actual field testing. The design solution will be fit to actual workers in the design environment and tested during the task itself. This may involve real-world situations as have been used in testing protective clothing for pesticide applicators such as that reported by Nigg et al. [9]. The field evaluations include objective evaluations of the design specifications, for example, measuring the penetration of pesticides through the use of pad techiques. Also, the evaluation of the prototype should include the users' evaluations of the satisfactory performance of the garment. Therefore, subjective user evaluations should be included as a complement to the objective evaluation.

After the prototype evaluation is complete, the designer takes the results of both the objective and subjective evaluations and incorporates these into the final design. The functional design process is never complete; it is an ongoing process, and with the execution of the final design the designer should be open to suggestions for improvement. Or if the designer becomes aware of problems with the final design in actual use, the process may begin all over again.

REFERENCES

1. S. M. Watkins, *Clothing the Portable Environment*, Iowa State University Press, Ames, Iowa, 1984.
2. J. C. Jones, *Design Methods*, John Wiley, New York, 1970.
3. J. O. DeJonge, E. P. Easter, K. K. Leonas, and R. M. King, Protective apparel research, *Dermal Exposure Related to Pesticide Use, ACS 273*, American Chemical Society, Washington, D.C., 1985, pp. 403–411.
4. J. O. DeJonge, G. Ayers, and D. Branson, Pesticide deposition patterns on garments during air blast field spraying, *Home Economics Research Journal 14* (2):262–268 (1988).
5. R. Fenske, Use of fluorescent tracers and video imaging technology to evaluate chemical protective clothing during pesticide applications, *Performance of Protective Clothing: Second Symposium ASTM STP 989*, American Society for Testing and Materials, Philadelphia, 1988, pp. 630–639.
6. W. F. Durham and H. R. Wolfe, Measurement of the exposure of workers to pesticides, *Bulletin of World Health Organization 75* (26):75 (1962).
7. H. N. Nigg, J. H. Stamper and R. M. Queen, Dicofol exposure to Florida citrus applicators: Effects of protective clothing, *Archives of Environmental Contamination and Toxicology 15:*121 (1986).
8. D. H. Branson, J. O. DeJonge, and D. Munson, Thermal response associated with prototype pesticide protective clothing, *Textile Research Journal 56* (1):27 (1986).
9. H. N. Nigg, J. H. Stamper, E. P. Easter, W. D. Mahon, and J. O. DeJonge, Protection afforded citrus pesticide applicators by coveralls, *Archives of Environmental Contamination Toxicology 19:*635 (1990).
10. C. A. Dimit, E. P. Easter, and J. O. DeJonge, The effect of seams and closures on pesticide penetration through fabrics, *Abstract Booklet: Fourth International Symposium on the Performance of Protective Clothing for the 1990's and Beyond*, sponsored by ASTM Committee F-23, Montreal, Quebec, Canada, 1991, p. 58.
11. R. Mull and J. F. McCarthy, Guidelines for conducting mixer/loader-applicator studies, *Vet. Hum. Toxicol. 28* (4):328 (1986).
12. U.S. EPA, Pesticide assessment guidelines subdivision U—applicator, exposure monitoring. EPA PB87-133286, Washington, D.C., 1987.
13. World Health Organization, Field surveys of exposure to pesticides: Standard protocol, GIFAP Tech Mono 7, VBC/82.1, Geneva Switzerland, 1982.

3
Chemical Protective Clothing

Mastura Raheel University of Illinois at Urbana-Champaign, Urbana, Illinois

I. INTRODUCTION

Occupational exposure of the skin to toxic chemicals is a recognized health problem. In the United States dermatitis is the major occupational disease [1–4]. The Bureau of Labor Statistics (BLS) reported that, for 1980, skin diseases accounted for 43% of all occupational illnesses reported by the private sector [5]. These data probably are an underestimation of the problem since only cases that received medical treatment are reported. The important concern is the potential for system damage due to absorption though the skin of a toxic chemical. The effects can range from systemic intoxication to promotion of cancer.

It is estimated that more than 100,000 chemical products with very different toxicological properties are in use throughout the world [6]. During the last decade, a number of large-scale accidents in chemical plants occurred, and many contaminated dumps hidden inside city boundaries were discovered [7]. Also, a number of workers became disabled due to chronic intoxication from various chemicals. However, as knowledge of risks from hazardous chemicals has become more recognized, the use of chemical protective clothing and equipment also has increased. This is due in part to regulations and standards requiring the use of chemical protective clothing (CPC) and equipment. Also, it is well documented that the use of CPC is cost effective compared to installing engineering controls, which may not work for all situations related to toxic exposure.

The majority of workers with a potential for skin contact by hazardous materials are protected by utilizing some form of chemical-resistant clothing, ranging from everyday clothing in conjunction with rubber gloves and boots to totally encapsulated suits. An appropriate level of protection in any particular situation is related to:

1. The potential effects of skin contact with the chemical
2. The exposure period
3. The body zone of potential contact
4. The permeability or penetration potential of the protective garment or item
5. The characteristics of potential contact, for example particulate, liquid, or gaseous
6. The additive or synergistic effects of other routes of exposure, for example, inhalation and ingestion
7. The physical properties required of the protective garment or item such as flexibility, puncture and abrasion resistance, and thermal protection
8. Cost

Presently, only limited data are available for objective evaluation of the toxic potential of skin contact. This is especially critical since toxic potential should be the major criterion for selection of protective garments or gloves [6]. Dermal toxicity assessments have been constrained due to in vivo testing of permeability of most chemicals through intact skin, lack of knowledge regarding systemic toxicity of chemicals permeating the skin, and difficulty in applying laboratory data generated through the ASTM permeation method to the human model. A few skin exposure studies have been conducted for some organic solvents such as benzene and selected pesticides [8,9]. However, animal studies on hairless mice and pig skin, as well as rate-controlling studies of dermal drug delivery, are expected to furnish much information on toxic potential of skin contact with chemicals [6].

Direct exposure to toxic chemicals may occur during both routine and emergency chemical handling. Chemical hazards from particulate dusts, liquids (sprays), or gases may be encountered by personnel in both civilian and military sectors. Examples occur in the manufacturing of fertilizers, pesticides, pharmaceuticals, and petrochemicals; maintenance and quality control activities associated with chemical processes; chemical waste handling; emergency response to chemical spills or fires; and chemical warfare.

Selection of chemical protective clothing is a complex task, and the consequences of making wrong selections can vary from a skin rash to a life-threatening situation. Indeed, a totally impermeable garment that is indestructible by any chemical with toxic properties would be the answer. However, even if such a garment and protective gear were commercially available, cost and human comfort factors would determine the use of such chemical protective clothing.

Considerations for the development of personal protection systems against hazardous chemical should therefore include problem definition, the choice of materials in developing protective clothing, methods for assessing chemical resistance, decontamination, physical properties of protective materials, as well as objective and subjective human factors.

II. CHEMICAL PROTECTIVE CLOTHING MATERIALS

New materials that can be made into items of chemical protective clothing have been developed by polymer chemists and material scientists. These materials offer superior performance to what was achievable in the past. Not only are materials produced that have excellent chemical resistance, but manufacturers also are able to combine these materials in a variety of ways that result in composites. The overall performance of the composites is generally better than might be predicted from the performance characteristics of the component materials [10].

Chemical protective clothing is intended to act as a barrier between the worker and the chemical hazard. Items within CPC range from goggles to gloves to full-body encapsulating ensembles. Respirators are part of personal protective equipment (PPE) but are not discussed here within the context of CPC. The items of CPC generally are grouped in five classes that represent either specific parts of the body for which protection is required or specific exposure conditions.

The CPC items that protect specific body areas are

1. Hood, face shield, and goggles for head, face, and eyes, respectively
2. Gloves and sleeves for hands and arms
3. Coat, jacket, pants, apron, and bib overalls for partial torso
4. Coveralls and full-body encapsulating suit for complete torso
5. Boots and shoe covers for feet

Figures 1–5 show several of these items as components of the protective ensemble. Face shields are generally made of clear acetate, cellulose propionate, polyester, polyvinyl chloride (PVC), or polycarbonate plastic. All other items of CPC may be made of one or the other polymer, as a supported or coated fabric or a multilayer composite. For example, gloves may be supported or unsupported, and made of natural rubber, butyl rubber, neoprene,

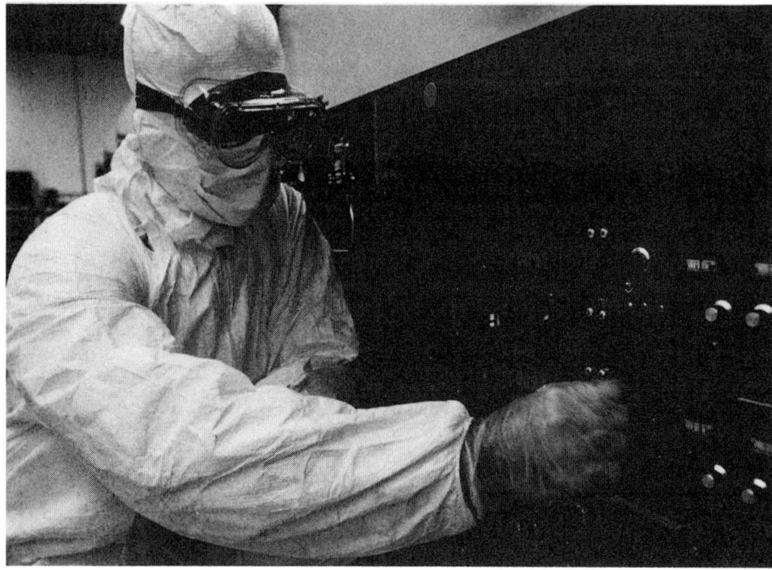

Figure 1 Clean room garment of Tyvek provides two-way protection from particles down to 0.5 μm size. Also shown are protective goggles and gloves. (Courtesy DuPont Tyvek.)

nitrile, polyvinyl alcohol (PVA), PVC, polyethylene (PE), Teflon-polytetrafluoroethylene (PTFE), Viton, or a composite of many materials. These same materials and several others are used for CPC suits as pure films, coated fabrics, laminates, or composites.

The primary requirement of chemical protective suit and glove materials is that they provide a protective barrier against toxic chemicals. These materials vary considerably in their ability to resist permeation and penetration of chemicals [11]. They also possess a wide range of thermal properties.

Tyvek, a spunbonded polyethylene, is a nonwoven uncoated or coated fabric used for disposable protective clothing. Uncoated Tyvek is a good barrier to many dry particulates including asbestos, radioactive dusts, and other hazardous particles. Polyethylene-coated Tyvek offers splash protection against acids and some other hazardous liquid chemicals. Tyvek laminated with Saranex forms an effective barrier against the permeation of many toxic chemicals including polychlorinated biphenyls (PCBs). Saranex, used as a film, is a multilayer laminate of polyethylene and Saran. Saran is a copolymer, made by polymerizing vinylidene chloride and vinyl chloride.

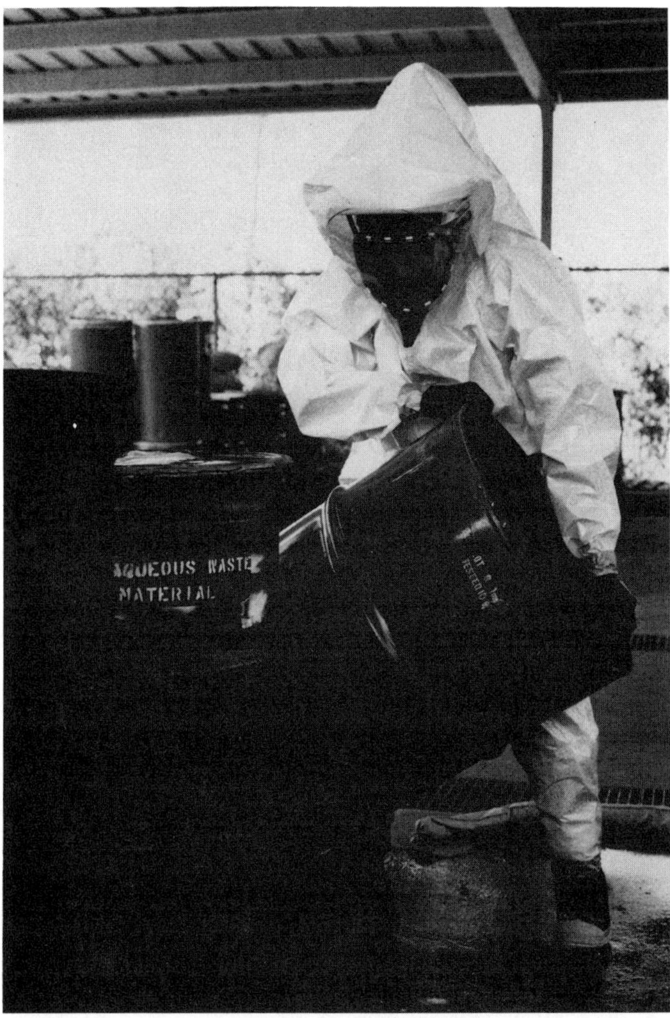

Figure 2. Laminate Tyvek/Saranex 23-P forms an effective barrier against the permeation of many toxic chemicals. Hazardous waste disposal personnel's ensemble shows protective suit, face shield, gloves, and boots. (Courtesy DuPont Tyvek.)

Figure 3. Garment of Tyvek laminate provides protection during pesticide mixing, loading, and application. Also shown are protective goggles, dual cartridge half mask respirator, hat, gloves, and boots. (Courtesy DuPont Tyvek.)

Tyvek, spunbonded olefin, is a nonwoven uncoated fabric used in disposable clothing. A newer material Comfort-Guard II, which is a sandwich construction with a breathable barrier coating in the middle, has been developed for limited-use garments. The outer layer is nonwoven olefin fiber, the breathable coating is PTFE polymer, and the inner layer is scrim. The composite fabric

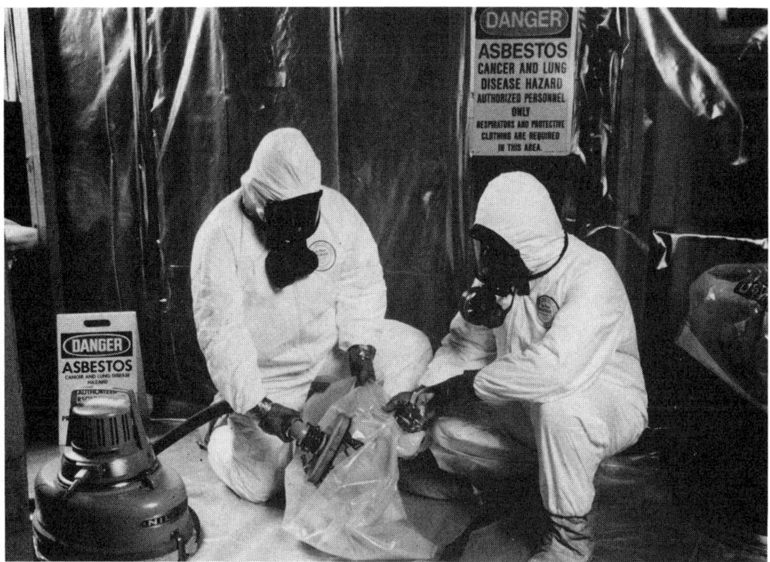

Figure 4. Protective one-piece garment of coated or laminated Tyvek used against particulate hazards. Asbestos abatement crew in suits with supplied-air respirators. (Courtesy DuPont Tyvek.)

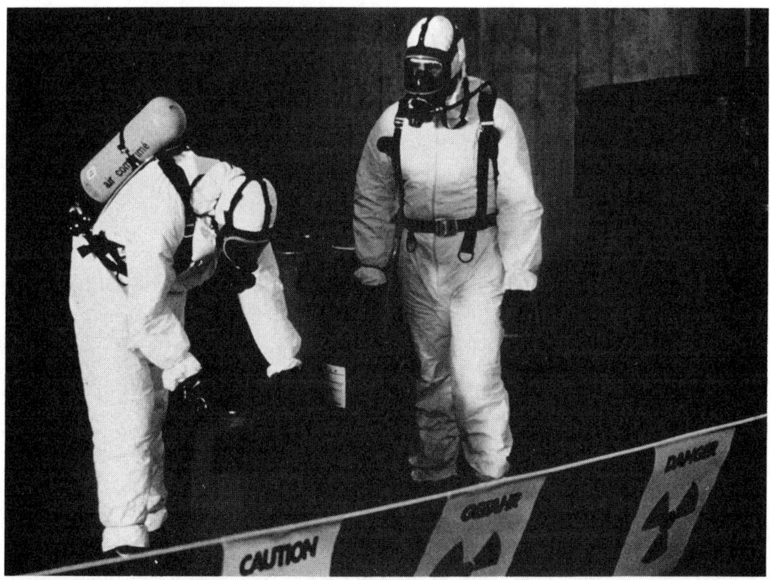

Figure 5. Protective ensemble in nuclear environments. Includes airsuit, helmet, respirator, self-contained breathing apparatus, gloves, boots, etc. (Courtesy DuPont Tyvek.)

garment is claimed to be as comfortable as cotton coveralls yet provide protection against particulates, general waterborne splashes, salt solutions, irritants, defatting agents, and some pesticides.

Polyvinyl chloride is a stiff polymer that is made suitable for multiple-use (durable) protective clothing applications by the addition of plasticizers. It is used as a film or as a coating over a base fabric and provides chemical splash protection. Butyl rubber also is used for the same purpose. Other materials have been developed to provide chemical protection as well as protection from heat exposure. These materials use heat-resistant fabrics, such as fiberglass or Nomex, laminated with Teflon PTFE chemical barrier film. Teflon PTFE material is chemically inert to almost all industrial chemicals and solvents. Teflon tetrafluoroethylene (TFE) resin is an opaque, white material. When softened, it does not flow like other thermoplastics and must be shaped initially. Maximum continuous service temperature is 260°C (500°F). Gortex PTFE also is a chemically inert film but is air permeable; therefore, it is more comfortable. It is used as a sandwich film between fabric layers. Caution is recommended when using these fabrics against splash protection, since gases and volatile organic liquids may permeate through them. Viton is another highly engineered material that has been introduced as a primary material for durable protective suits and gloves for handling chlorinated hydrocarbons including PCBs and aliphatic and aromatic hydrocarbons. Polymers of Viton are manufactured by polymerizing combinations of vinylidene fluoride (VF_2), hexafluoropropylene (HFP), and TFE. Viton has a unique balance of physical properties. It is reportedly superior to any other elastomeric material in chemical resistance to aromatic and chlorinated solvents and to degradation at elevated temperatures [12].

A. Protection Against Particulate Hazard

Hazardous airborne particulates exist in many industrial and military situations where personnel wear permeable clothing. The resistance to attraction, penetration, and retention of hazardous particles from airborne dust largely determines the effectiveness of protective materials. It has been suggested [13] that the static charge that develops on synthetic fiber fabrics with high dielectric properties increases the level of protection that they afford against finely dispersed aerosols. For example, in the beryllium industry where work garments are made of Nomex aramid fiber, it was found that Nomex fiber garments retained toxic particles more than cotton fiber garments and did not resuspend a large fraction of its contaminant load. Another case in point is protection against asbestos. Coated or laminated Tyvek is reported to have good barrier properties against asbestos [14,15]. Tyvek is a polyethylene nonwoven fabric of DuPont Co. recommended for disposable protective clothing. The product provides adequate protection against particulate contaminants such as asbestos down to a size of 0.5 mm. Polychlorinated biphenyls are another chemical hazard that firefighters or power and railroad workers who service capacitors and transformers containing PCB's [15] may encounter. Stampler et al. [16] have studied the permeation of PCBs through selected protective clothing materials and recommend nitrile rubber, Viton, Viton SF, and Vitrile as glove materials that provide best protection against PCBs. Also, garments made from Tyvek coated with polyethylene or laminated to the Saranex film (Dow chemical) are specified against PCBs [15].

Saranex-laminated Tyvek and polyethylene-coated Tyvek also are nonpermeable toward airborne dye and drug dust. Airborne dye and drug dust may be generated in storage, movement, weighing out, and mixing in powder form. The potential inhalation of dye dust, particularly from reactive dyes in powder form is an occupational hygiene problem. Along with the engineering controls (exhaust ventilation and easy-to-clean surfaces), respiratory protection and high-efficiency ventilated visors or airstream helmets also are recommended.

The permeability of disposable protective clothing materials to antineoplastic drugs has shown that Saranex-laminated Tyvek is not permeable and that polyethylene-coated Tyvek is slightly permeable to selected drugs. Noncoated Tyvek was found to be permeable to fine-particulate drugs. Clothing made from the two Tyvek composites would allow less airflow and would therefore be less comfortable to wear for extended periods. Garments made from noncoated Tyvek would be more comfortable, but their use should be accompanied by an awareness of their potential permeability to certain antineoplastic drugs, dyes, and other fine-particulate hazards.

B. Protection Against Liquid Hazards

The most common cause of injury among chemical workers is penetration of liquid chemicals through their clothing due to spillage. Both chemical resistance and impermeability of the protective garment are desirable. This would, of course, depend upon the target chemical(s) and exposure period. Hence, different fibers, fabrics, or composites are recommended in different chemical industries along with garment design and construction features. For protection against acid and other corrosive chemicals, for example, wool fabrics treated with Zirpro flame retardants as well as water- and oil-repellent fluorochemical finish have exhibited good barrier, mechanical and hygienic properties [17]. Double-faced fabrics produced in wool/polypropylene-fiber offer similar protection to that of fluorochemical-finished single wool fabrics against most chemicals, particularly against nitric acid. Benisek [18] has shown that the effectiveness of a protective finishing treatment depends not only on the chemical composition and concentration of the finishing agent but also on the chemical composition and physical parameters of the fabric (weave, thickness, etc.). Table 1 presents the resistance of these fabrics to penetration by selected liquids.

Protective work clothing for operatives dealing with petroleum and its products and those who clean out storage tanks must be impermeable to petroleum products and be heat and fire resistant. Fluorochemical-finished wool and aramid-fiber fabric were tested in a simulated gasoline-bomb test. The test involves pouring gasoline on the fabric specimen—which rests at an angle of 57° to the horizontal with a simulated PVC skin beneath it—igniting the fabric, and evaluating the damage to the skin. Results showed much greater damage to PVC skin under aramid-fiber fabric than fluorochemical-finished wool fabric. This is due to the oleopilic nature of aramid fiber. Tyvek spunbonded fabric surface-coated with polyethylene or laminated with Sar-

Table 1 Resistance to Penetration of Chemicals, Gutter Test[a]

Fabric description	Aqueous wetter, 35 dyn/cm	Sodium hydroxide, 75%	Acids[b] HCI, 20%	HNO_3, 70%	H_2SO_4, 98%
Woolen fabric, 600 g/m^2, 1.1 mm thick	Pass	Pass	Pass	Fail	Pass
Woolen fabric, 600 g/m^2 2.1 mm thick	Pass	Pass	Pass	Fail	Fail
Worsted fabric, 270 g/m^2, 0.7 mm thick, 2/1 twill	Fail	Pass	Fail	Fail	Fail
Worsted fabric, 270 g/m^2, 0.7 mm thick, 2/1 twill (water-repellent-treated)	Pass	Pass	Pass	Fail	Pass
50/50 wool/polypropylene-fiber double cloth, 420 g/m^2, 1 mm thick, wool face exposed	Pass	Pass	Pass	Pass	Pass
Polypropylene-fiber face exposed	Pass	Pass	Pass	Pass	Pass

[a]The gutter test, draft DIN 32763, measures the penetration of liquids through a fabric.
[b]Pass indicates no penetration of chemical.
Source: Adapted from Ref. 17.

Table 2 Permeation Resistance of Selected Protective Clothing Materials

Material description	Chemical	BT	SSPR
Viton/nylon/ chlorobutyl rubber	Acetone	29	196
Chlorinated polyethylene	Trichloroethylene	96	8050
Saranex/Tyvek	Diethylamine	21	368
Teflon/Nomex/Teflon	Dichloromethane	32	0.32

Abbreviations: BT, permeation breakthrough time in minutes; SSPR, Steady-state permeation rate ($\mu g/cm^2$-hr).
Source: Adapted from Ref. 19.

anex film offers good protection against gasoline spills but not against heat or flames. The permeation resistance of selected protective clothing materials to some target liquid chemicals has been reported by An et al. [19] as shown in Table 2.

Schwope et al. [20] have conducted extensive studies on glove materials' permeation to agricultural pesticides. Identifying appropriate handwear is of critical importance, since it has been shown that 87% of dermal exposure to pesticides occurred through the hands [70]. Schwope et al. evaluated the permeation resistance of seven generic types of glove materials toward 20 pesticide formulations, which were generally solvent-based concentrates. Table 3 lists the active ingredients and carrier solvents present in these formulations. Seven generic types of glove materials used in their study are listed in Table 4. The researchers concluded that the carrier solvent generally permeated sooner and at a much higher rate than the active ingredient. Also, the better glove materials (among those investigated) were nitrile rubber, butyl rubber, and plastic film laminates (Silver Shield). Natural rubber, polyethylene, and polyvinyl chloride did not provide adequate protection [20].

C. Protection Against Toxic Fumes and Gases

Exposure to vapors penetrating protective clothing is, in general, negligible when compared to inhalation exposures for low vapor concentration. Thus, safety masks for inhalation protection cannot be overemphasized. In situations where high concentrations of hazardous gases or fumes may be encountered, a fully encapsulated suit with self-contained breathing apparatus is recommended. However, under low vapor concentration situations, neoprene, butyl rubber, polyvinyl chloride, and polyethylene garments in con-

Table 3 Active Ingredients and Carrier Solvents Present in the Pesticide Formulations

Component	Types of compounds
Active ingredients	Dicrotophos
	Endosulfan
	Ethion
	Ethyl parathion
	Methyl parathion
	Mevinphos
	Momocrotophos
	Naled
	Oxydemeton
	Phosalone
Carrier solvents	
Alcohols	Isopropanol
	Hexylene glycol
Ketones	Acetone
	Cyclohexanone
	Methyl isobutyl ketone
Petroleum distillates, aliphatic	Kerosene
	Mineral oil
	Petroleum oil
Petroleum distillates, aromatic	Xylenes
	Xylene range solvents

Source: Ref. 20.

junction with face masks are adequate. The use of activated carbon particles in protective fabrics reduces diffusion of gases, fumes, as well as liquids. Microporous polymeric matrices with active particulate carbon fixed throughout the pore structure have been developed. These matrices include solid and hollow textile fibers such as polyolefin fibers containing activated carbon, as well as carbon-loaded films. These textiles are generally in the form of composites such as bimembranes with an encapsulated active-carbon layer or a composite fabric with activated carbon-impregnated polyurethane foam sandwiches between two fabric layers.

Activated fiber, yarn, or fabric can be prepared by pyrolysis of precursors such as rayon, or polyacrylonitryl or phenolic (kynol) fiber to carbon in an inert atmosphere [21]. Pyrolysis is followed by activation (pore and surface area formation) at a high temperature in an oxidizing gas atmosphere (e.g.,

Table 4 Glove Materials Tested

Glove type	Source	Model	Average thickness, cm
Butyl rubber	North	Butyl B174	0.05
	Mine Safety	#38553	0.04
Natural rubber	Granet	541	0.05
	Pioneer	Ivory White	0.06
Neoprene	Edmont	Neoprene 29-870	0.05
	Ansell	Neoprene	0.05
Nitrile rubber	Edmont	Sol-Vex 37-155	0.04
	Pioneer	Stansolv A-15	0.04
	Ansell	Challenger 613	0.04
	Best	Nitrile #727	0.04
Polyethylene	Fisher	Polyethylene	0.01
Polyvinyl chloride	Pioneer	Stan-flex	0.05
Silver Shield	North	Silver Shield	0.01

Source: Ref. 20.

carbon dioxide or super-heated steam). Adsorption studies have shown that activated carbon fibers have pore radii between 5 and 14 Å and display excellent adsorption capabilities for liquids and gas pollutants such as phenol, butane, HCN, H_2S, and NO_2.

The same basic process is used in the manufacture of granular activated carbon. One such example is Ambersorb (a trademark of Rohn and Haas Co.) carbonaceous adsorbent spheres. These adsorbent beads have been used in protective garments made of Helsasafe (a trademark of Helsa-Werke, Helmut Sandler GmbH and Co. of Germany) fabrics. These fabrics were originally developed for protection against chemical warfare agents and were found to perform effectively. Helsasafe fabrics may be fully laminated composites. One such fabric is Helsasafe 2032 in which the outer layer of viscose tricot is bonded to the center layer of Amersorb 572 beads, which in turn is bonded to the inner layer of cotton fabric. In another fabric type, the outer fabric of polyester/cotton (#503045) or cotton/glass (#3502666) is not bonded to the center adsorbent layer, which, however, is bonded to the inner layer of flame-retardant cotton fabric. Garments made of the latter type of fabric are less stiff than the fully laminated fabrics. Head hoods containing activated carbon for protection against toxic fumes from burning buildings or vehicles also have been developed.

III. BARRIER EFFECTIVENESS OF CHEMICAL PROTECTIVE CLOTHING: THEORETICAL BASIS

The Toxic Substances Control Act (Public Law 94-469) in its section 5 requires prospective manufacturers of chemicals to submit Premanufacture Notifications (PMNs), which are reviewed by the U.S. Environmental Protection Agency (EPA) Office of Toxic Substances (OTS) prior to the manufacture or import of new chemicals. Even though many substances are not subjected to all aspects of the review process, those that are judged potentially toxic require detailed assessments of the potential for their environmental release and human exposure during manufacture, processing, and end use. If concerns are raised to warrant regulation, then engineering controls, work practice restrictions, and protective clothing and equipment are investigated as a means to reduce exposure risks [22].

The prospective manufacturer of a chemical often recommends protective clothing as a means to limit dermal exposure. Occasionally, the type of clothing is specified; more often it is not. In either case, Office of Toxic Substances must have a means of assessing the exposure reduction provided by protective clothing. This can be achieved by actually testing the clothing materials or by an estimation of clothing performance using available information and predictive models.

The overall assessment of protective clothing requirements must consider the potential health effects of the chemical, the probable exposure conditions, and the effectiveness of clothing in limiting exposure. The barrier effectiveness of chemical protective clothing can be measured by permeation testing. The objective of conducting standard permeation tests is to obtain measurements that allow direct comparison of different protective clothing materials and challenge chemicals and that are representative of the properties of the materials in actual use. Clearly, this objective is compromised by variation in both test procedures and conditions of protective clothing use.

In order to control variability in test procedures, the American Society of Testing and Materials (ASTM) Committee F-23 on Protective Clothing established the Test Method for Resistance of Protective Clothing Materials to Permeation by Liquids or Gases (F 739-85). The method quantifies the permeation of liquids through protective clothing materials under conditions of continuous contact [23]. Consequently, the method has been widely used for obtaining chemical resistance data for materials used under a range of field conditions. Despite the use of a standard test method, however, comparing test results and applying them to the field remain difficult. This has been particularly true with the measurement of breakthrough time, which, although commonly used, is extremely sensitive to variations in analytical detection limits of the test method. An understanding of the theoretical basis of these test methods is essential for accurately interpreting and comparing CPC test results for real-world situations.

A. Permeation Theory

Permeation is the molecular process by which chemicals move through protective clothing materials. The mechanism of permeation involves three steps: (1) absorption of individual molecules of the chemical into the exposed surface of the material, (2) molecular diffusion through the material matrix along a concentration gradient, and (3) desorption of the chemical from the inside surface [24]. The mathematical formulation of molecular diffusion through a plane sheet of material has been described by Crank [25], and application to chemical permeation of polymeric materials has been reported by Bomberger et al. [26], Schwope et al., [27], and Coletta et al. (28). The mathematical theory of diffusion based on Fick's first law states that the permeation rate of a chemical through an area of material is proportional to its concentration gradient through the material. The proportionality constant D is the diffusion coefficient, which may be a function of chemical concentration and therefore time. In many cases, however, D is treated as a constant. If D is constant, the one-dimensional differential equation of diffusion, Fick's second law can be expressed as

$$\frac{\partial c}{\partial t} = D \frac{\partial^2 c}{\partial x^2}$$

where c is the concentration of the diffusing chemical. This equation is derived from Fick's first law by considering diffusion through a volume element.

If it is assumed that chemical permeation through protective clothing material obeys Fick's laws, the diffusion equation can be solved to obtain the permeation rate and chemical mass permeated as a function of time. The following assumptions and conditions are applied:

1. The diffusion coefficient D is constant (independent of solvent concentration in material).
2. Swelling of the material and degradation by the permeating chemical are negligible.
3. At time zero, one surface is exposed to the chemical and immediately reaches a constant concentration.
4. The concentration of chemical on the unexposed side of the material is kept negligible compared to that of the exposed side.

Solving the above differential equation by Laplace transformation techniques, the total amount of chemical that has permeated a unit area of the material at time t can be expressed as

$$m_t = \frac{\overline{D}c_1 t}{a} + \frac{2c_1 a}{\pi^2} \sum_{n=1}^{\infty} \frac{(-1)^n}{n^2} \left[1 - \exp\left(-\overline{D}n^2\pi^2 t/a^2\right)\right]$$

where

c_1 = concentration of chemical on the exposed side,
a = material thickness, and
$\overline{D}$ = assumed constant diffusion coefficient (integral value of the diffusion coefficient over the range of chemical concentrations encountered).

At large values of t (after long exposure), the exponential terms in the equation become small, such that the amount of permeated chemical versus time can be expressed by

$$m_t \approx \frac{\overline{D}c_1 t}{a} = \frac{c_1 a}{6}$$

This is the permeation equation at steady state where the total amount of chemical permeation increases linearly with time (constant permeation rate). From the equations for m_t, the concentration of challenge chemical in the collection side of a closed-loop permeation test cell can be expressed as a function of time by

$$c_t = \frac{m_t A}{V}$$

where A is the cross-sectional area of the exposed material and V is the volume of the collection side of the cell.

An equation for the permeation rate of the chemical J_t, as a function of time, per unit area of material also may be obtained from the diffusion equation

$$J_t = \frac{\overline{D}c_1 t}{a} \left[1 + 2\sum_{n=1}^{\infty} (-1)^n \exp\left(-\overline{D}n^2\pi t/a^2\right)\right]$$

At large t, the exponential terms vanish, leaving a constant (steady-state) permeation rate.

The concentration of chemical on the collection side of an open-loop permeation test cell is expressed by

$$c_t = \frac{J_t A}{F}$$

where F is the flow rate on the collection side of the cell. The chemical concentration on the collection side of an open-loop system at steady state is obtained by substituting the steady-state expression for J_t, giving

$$c_t \approx \frac{\overline{D}cA}{aF}$$

The breakthrough time of a chemical through a material, as measured by the standard permeation test method, can be modeled by the preceding permeation equations. It is defined as the time t at which the concentration c_t is detected in the collection side of the cell. If detection occurs prior to the beginning of steady-state permeation, breakthrough time occurs both in the linear term and in the exponential term within the infinite series of the permeation equation. Thus, no simple function can be derived relating breakthrough time of a given chemical/material pair to the variables of material thickness, detector sensitivity, or cell size.

Several researchers have shown the importance of considering, controlling, and reporting the experimental conditions under which permeation tests are conducted [29]. This is critical not only in accurately interpreting and comparing results but in applying test results to real-world situations [24]. Chemical/material combinations that show no detected breakthrough in standard 3-hr tests may in fact permeate within that time under elevated temperatures or if the material used is actually thinner than the rest specimens.

Temperature has a quasi-logarithmic effect on breakthrough time, indicating the need for close temperature control and temperature reporting for all permeation tests. In addition, permeation data should be obtained, or results predicted, for the full range of temperatures that a material might encounter in use. Also, an apparent quadratic increase of breakthrough time with material thickness has been reported [24].

Although breakthrough time is commonly used, steady-state permeation rate is perhaps a better measurement for comparing test results. The permeation rate J_t at steady state is a simple inverse function of material thickness and, if measurable, is independent of cell size and analytical detection limit. However, limitations to the practical use of steady-state permeation rate are the possible long experimental times required to reach steady state and the large chemical concentrations that sometimes must be measured. Also, information important in protective clothing use, such as the time required for accumulation of an appreciable amount of permeated chemical, is not always provided. Dangerous amounts of chemical may have permeated well before steady-state permeation has been reached. In such cases, breakthrough time of mass permeated by a given time is a better measure of permeation.

B. Solubility and Diffusion Parameters

The concentration of permeant c, within the solid material (CPC) is estimated using Henry's law:

$$C = S_p$$

where S is Henry's law constant or solubility coefficient and p is pressure at the upstream and downstream boundaries [30]. This is used in the solution of Fick's first law to obtain flow rate F:

$$F = DS(p_1 - p_2)/1$$

permeation coefficient P can be defined as

$$P = DS$$

Thus,

$$F = P(p_1 - p_2)/1$$

where P, D, and S are determined experimentally through permeation testing.

Two types of interaction between the permeant(s) and the material are important in permeation. The first type of interaction is the solubility of the permeant in the matrix of the material, whereas the second is the mobility of the permeant molecules in that matrix.

1. *Solubility*

The solubility of the permeant in the solid describes the ease with which the permeant molecules go into solution within the matrix of the material. Solubility parameter is determined by estimating Gibbs Free Energy of Mixing [31–34].

Hildebrand defined the solubility parameter for a liquid as the square root of the cohesive energy density of the liquid, which is equal to the molar change in internal energy on vaporization per mole volume. This total solubility parameter, the Hildebrand parameter, is used to predict whether two materials will form a solution. If the total change in internal energy (ΔE) is small, the two should mix freely.

ΔE can be determined for solvents. Since polymers degrade before vaporizing, ΔE for polymers can be determined only indirectly, by determining the swelling behavior of a polymer in solvents of known solubility parameter.

Hansen [33] extended Hildebrand's concept by proposing a multifactor solubility parameter, made up of energy terms for each of the four types of molecular interaction that occur. These are polar, including both permanent and induced dipoles, hydrogen-bonding, and dispersion forces. The three terms are squared and summed, yielding the total solubility parameter described by Hildebrand.

Hansen's three-dimensional solubility parameters are combined for mixtures of solvents in the same fashion as the total solubility parameter. The parameters have been applied successfully to the selection of solvents in the paint and coatings industry and in a number of other fields [30]. The three-dimensional solubility parameter should predict the relative ease of the solution step for mixtures, perhaps better than it predicts the extent of the equilibrium solubility.

Morris and Wagner [35] studied nitrile rubbers with 22 to 40% acrylonitrile content. They showed that increasing the nitrile content decreased the swelling caused by absorption of a mixture of toluene and *i*-octane. Morris and Wagner determined that toluene was preferentially absorbed from the mixture and that there was an optimum toluene:*i*-octane ratio for maximum swelling and maximum *i*-octane absorption. They related the preferential absorption of toluene over *i*-octane to the greater similarity between the cohesive energy densities of nitrile rubber and toluene than nitrile rubber and *i*-octane. More recently, Abu-Isa [36,37] and Myers and Abu-Isa [38] have reported that swelling of various elastomers by a mixture of gasoline and alcohols correlates well with three-dimensional solubility parameters.

2. *Diffusion*

The diffusion of a permeant is related to the ease with which a permeant molecule "fits" within the lattice of polymer chain segments and is able to move from place to place within that lattice. The frequency with which diffusion takes place depends on three factors: the size and shape of the diffusing molecule, the tightness of packing and intermolecular forces between polymer chains, and the stiffness of the polymer chains [39,40]. Membrane swelling, crystallinity, and temperature affect permeability and permeation selectivity by altering these three factors. For example, swelling loosens up the forces between polymer chains and relaxes chain stiffness. This reduces the energy required for a permeant molecule to move from one to another open space and increases the number and size of open spaces. Hence, there is an increase in permeability and decrease in permeation selectivity.

Researchers studying the effects of mixtures on CPC have noted that different components do not permeate at the same rate [38,41,42]. Others have postulated that one solvent "carries" another through the elastomer [43,44]. This is probably due to the swelling effect.

3. Persistent Permeation

After a chemical has begun to diffuse into a plastic/elastomer, it will continue to diffuse even after the chemical on the surface is removed. This is due to the concentration gradient that develops across the CPC and the natural tendency for a gradient to equilibrate with its surroundings. The phenomenon has significant implications relative to decontamination and reuse of CPC [45]. For example, in the case of CPC, which has not suffered chemical breakthrough but has absorbed a certain quantity of a chemical before the chemical is removed from the surface, the chemical may eventually appear on its inside surface. The amount of chemical reaching the inside will be dependent upon the amount of chemical absorbed and its permeation rate. Thus, in order to achieve complete decontamination of the CPC, both surface and absorbed chemical must be removed. Since the absorbed chemical will leave the CPC only by a diffusional process, either very long time or conditions that accelerate diffusion are required. These would include high temperature, vacuum, or perhaps a dry-cleaning process in which a chemical nondegrading to the CPC is used to extract the hazardous chemical. Because of the problem of persistent permeation, extreme caution is advised when using CPC that has been exposed to highly toxic chemicals. In fact, where such chemicals are involved, it may be prudent practice to use disposable clothing [45].

IV. PERFORMANCE EVALUATION OF CHEMICAL PROTECTIVE CLOTHING

Chemical resistance, physical properties, and human comfort properties of a clothing material must be considered in judging its acceptability. Even though the physical property requirements are specific to the particular application, chemical resistance and human comfort are more general concepts.

A. Chemical Resistance of CPC

A comprehensive technical and reference manual in two volumes, *Guidelines for the Selection of Chemical Protective Clothing*, has been prepared by Arthur D. Little, Inc. [45]. This manual provides CPC recommendations for about 300 chemicals. For those chemical/material pairs for which no recommendations are given, it is suggested that CPC can be selected on the basis of the family to which the chemical belongs. The premise, which is substantiated in permeation literature, is that chemicals of similar composition tend to permeate a given material at relatively similar rates. Extensions and refinements of this premise follow:

1. Higher molecular weight members of a homologous series of chemicals permeate at slower rates than lower molecular weight members.
2. Pendant groups (which increase the size of a molecule) tend to slow the permeation rate relative to that of the simple molecule.
3. Permeation rate tends to decrease with increasing boiling point.
4. Polar chemicals tend to permeate polar materials more rapidly than nonpolar chemicals, and the converse is true.

The 300 chemicals have been categorized into 30 main classes and 52 subclasses according to structure and functional groups [45]. The class into which each chemical is placed can be determined from Appendix B and Users' Matrix A in Volume I. The generalizations stated appear to apply to chemical resistance of CPC for a large number of the chemical/material pairs [45]. Another important reference guide, *Chemical Protective Clothing Performance Index*, has been compiled by Forsberg et al. [46]. These researchers have compiled chemical breakthrough time and permeation rate data for about 600 chemicals and about 200 protective clothing models from reference sources all over the world and published it in a book. An electronic version of this book with updated information also is available and is distributed with the "GlovES+" Expert System to help find CPC with the best demonstrated performance.

A second means for predicting the chemical resistance of CPC materials is based on the solubility parameter described earlier. According to this theory, the physical and chemical properties of a chemical can be combined mathematically to yield a parameter that is then compared to an empirically determined parameter for the barrier material (CPC). In cases where the parameter of the chemical approximates that of the material, the chemical is predicted to have a high solubility in or dissolve the material. In other words, "likes dissolve likes." Extrapolation of this theory to CPC implies that a material is not likely to be resistant to a chemical having a similar solubility parameter.

The results of a limited number of tests of the theory relative to CPC materials would seem to support its application to CPC selection. However, many variations of this theory are found in the literature. Barton [47] and Snyder [48,49] have reviewed the factors that may affect the calculation of the solubility parameter. One of the more widely accepted concepts is Hensen's three-component parameter described earlier, which combines factors for hydrogen bonding, polarity, and dispersion forces of the chemical to yield its overall solubility parameter. Other systems deal with two of these factors. Still other systems favor the single-component solubility parameter then make adjustments for polarity or hydrogen bonding depending on the application. Solubility parameter theory offers great promise for predicting PC performance; however, the transfer of the theory to CPC is in its early stages and will require a significant level of technical evaluation before its applicability can be judged [45].

1. Test Methods for Chemical Resistance of CPC

The barrier effectiveness of a particular item of clothing to a particular chemical/mixture is dependent on the specific interactions between the clothing material and the chemical/mixture. This in turn is determined by the polymer type, fabrication method, finishes or coatings on the clothing material, and its thickness [50–54]. Temperature and other conditions of use such as perspiration [55], soiling [56], abrasion, or other physical damage also influence clothing barrier properties. Finally, the composition of the chemical/mixture is of major importance since relatively small percentages of a second, third, etc., component can drastically alter the way in which a chemical interacts with a material.

With the above in mind, it is highly desirable that decisions for selecting protective clothing be based on the results of testing of the chemical/clothing material pair of interest for both solubility and permeation.

(a) Solubility Tests. Solubility is the weight of chemical absorbed by a known weight of material. In general, chemicals having high solubility (i.e., $>10\%$) rapidly permeate the CPC material. The ASTM Method D471-79 and ISO (International Organization for Standards) Method 2025 describe methods for determining solubility. The procedure involves immersing the material in the chemical. In the case of multilayered clothing materials, only the surface normally outside should be exposed to chemical. Periodically the material is removed, patted dry, and weighed until a constant weight is obtained. In addition to noting weight changes, the chemical should be inspected for discoloration, indicative of decomposition of the clothing material. The clothing material should be examined for physical degradation using appropriate methods.

Solubility testing is simple and can readily be performed wherever at least a two decimal balance is available. However, it represents the minimum level of evaluation that can be performed for any unknown or multicomponent hazardous material.

(b) Permeation Tests. In a permeation test, the chemical of interest is placed on one side of the clothing material, and the other side is monitored for the appearance of the chemical. A schematic plan of the permeation cell is shown in Figure 6. From the results, the total amount (mass) of chemical permeating a known surface area of the clothing at any given time can be calculated. The cumulative permeation (mass/area) or the corresponding permeation rate (mass/area/time) can be used along with the estimated frequency, duration, and exposed body surface area to estimate dermal exposures for specific workplace activities.

Although the test is straightforward and a standard method (ASTM F 739-85) exists for its performance, as discussed earlier, variable results can be obtained under different testing conditions for the same chemical/clothing material. Consequently, in describing or interpreting the results of a permeation test, there is a certain minimum amount of information required. This information includes the breakthrough time, the steady-state permeation rate, the clothing material thickness and surface area, the analytical sensitivity, the

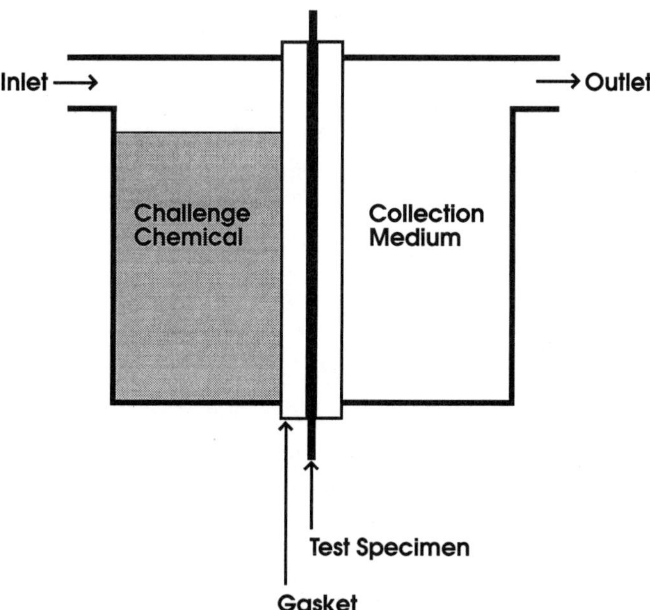

Figure 6. Schematic representation of ASTM F 739-85 permeation test cell.

collection medium flow rate (open-loop systems) or volume (closed-loop systems), and temperature [57].

Based on the types of results obtained from clothing performance tests, their approximate cost, the relative skill level required to perform them, and their inherent limitations, Goydan et al. [22] have recommended a testing hierarchy. The testing hierarchy (Table 5) ranks chemical-resistance tests according to their ability to generate data that can be used directly to estimate the effectiveness of the clothing material in reducing exposure risks.

B. Structural Integrity

In the selection and use of protective clothing, other factors, besides chemical resistance, may be of equal or greater importance. The factors include thermal and tactile properties of CPC, structural integrity, and human comfort aspects of the protective clothing systems. For example, gloves must provide the wearer some acceptable level of dexterity, be puncture and tear resistant, and provide chemical protection. Face shields and lenses, in addition to being chemical barriers, must provide clear, undistorted vision to the wearer. Hard, inflexible face shields and lenses may be subject to surface cracking (crazing) upon contact with certain chemicals. Crazing renders the surface foggy and can drastically reduce vision. Since chemical contact with the face shield or lens is more likely to occur in uncontrolled or emergency situations when reduced vision would be an additional source of hazard, shields and lens materials should be tested for resistance to chemical attack. Crazing can also reduce the impact strength of the material. The relative importance of the performance factors is largely dependent on the specific work tasks to be carried out.

Table 5 Test Data Priority for Estimating the Chemical Resistance of Protective Clothing Polymers

Chemicals	Test method(s)
Single-component liquids/gases	1. ASTM F 739-85 permeation test 2. Permeation cup (if chemical has sufficient vapor pressure) 3. Weight change from liquid immersion 4. D and S from vapor or liquid sorption/desorption test
Multicomponent solutions	1. ASTM F739-85 permeation test

Abbreviations: D, diffusion coefficient; S, solubility.
Source: Adapted from Ref. 22.

The ANSI/ASTM Method F484-77 describes a procedure for determining stress crazing by chemicals. Another method for determining the effect of chemicals on clear plastics is by measuring the transparency of the plastic before and after exposure to the chemical; ASTM D1746 describes one such method. At the present time, there is no standard or overall protocol for evaluating protective clothing or clothing materials for all the performance parameters of importance to workers exposed to hazardous environments. Instead, individual tests appropriate for the evaluation of specific parameters must be selected from the volumes of procedures promulgated by federal, military, and standards organizations. Table 6 shows some of the tests for physical property assessment of CPC [45], along with the test method number of the ASTM, Federal Standard, or reference information about experimental methods.

C. Human Comfort

Chemical resistance, barrier properties, and structural integrity of chemical protective clothing must be considered in relation to human comfort. Protective clothing is often made of impermeable fabrics that may in fact be a hazard in itself due to hyperthermia under conditions of high temperature, low wind velocity, and low evaporation rate [59]. In cold environments, workers can protect themselves by wearing heavier clothing, up to the point where the garments start to interfere with the performance of necessary skills. In addition, poor design of CPC sometimes contributes to deterioration in work performance and increased accident risks [60].

There are three international standards dealing with thermal effects of clothing on the wearer. These are ISO 7730, ISO 7243, and ISO DIS 7933 standards. These standards and their usefulness are described in detail by Olesen and Dukes-Dobos [61]. A summary of the standards is described here. The ISO 7730 standard describes a method for identifying conditions that make people feel thermally comfortable. It provides easily accessible information on the effect of clothing insulation on thermal comfort and includes tables presenting clothing insulation values for single garments as well as clothing ensembles. However, the validity of the standard is limited to conditions where the relative humidity is 50% and the clothing worn is water vapor permeable.

The ISO 7243 and ISO DIS 7933 standards describe methods for assessing the heat load of workers exposed to hot environments and provide safe exposure limits. The difference between the two standards is that ISO 7243 avoids the need for measuring clothing insulation, whereas ISO DIS 7933 provides equations to accommodate values of clothing insulation and water vapor permeability. As a consequence, the applicability of ISO 7243 is

Table 6 Structural Integrity Test Methods for Chemical Protective Clothing

A. *Visibility of face shields and lenses*	
Crazing	ASTM F484-77: Stress Crazing of Acrylic Plastics in Contact with Liquid or Semi-Liquid Compounds
Transparency	ASTM 1746-70: Transparency of Plastic Sheeting
Strength degradation	ASTM D543: Resistance of Plastics to Chemical Reagents
B. *Strength*	
Tear resistance and strength	ASTM D751-73: Testing of Coated Fabrics ASTM D412-75: Rubber Properties in Tension Fed. 191A-5102 (ASTM D1682): Strength and Elongation Breaking of Wove Cloth: Cut Strip Method Fed. 191A-5134 (ASTM D2261): Tearing Strength of Woven Fabrics by the Tongue Method
Puncture resistance	See Refs. 47 and 58
Abrasion resistance	ASTM D1175: Abrasion Resistance of Textile Fabrics
C. *Dexterity/flexibility*	
Dexterity (gloves only)	See Refs. 47 and 62
Flexibility	ASTM D 1388: Stiffness of Fabrics, Cantilever Test Method
D. *Aging Resistance*	
Ozone resistance	ASTM D3041-72: Coated Fabrics—Ozone Cracking in a Chamber ASTM D1149-64: Rubber Deterioration—Dynamic Ozone Cracking in a Chamber
U.V. resistance	ASTM G27: Operating Xenon-Arc Type Apparatus for Light Exposure of Non-Metallic Materials—Method A—Continuous Exposure to Light

limited to conditions where workers wear light and vapor-permeable garments, whereas ISO DIS 7933 can be used, in principle, for all combinations of clothing insulation and permeability.

Tables of insulation values for commonly worn garments are available, but such information is not available for protective garments. Values for the vapor permeability of these garments are also unavailable. As a result, ISO DIS 7933 has limited utility at the present time. The ISO Working Group for Thermal Environments is in the process of drafting yet another standard that will specify all the necessary measurements and calculations for assessing clothing insulation and water vapor permeability. The measurements and calculations for this purpose are complex, and the equipment, an electrically heated copper manikin, is expensive. Such manikins are available only in a few specialized laboratories in the United States and abroad. Since there is no simple and inexpensive methodology known at present for these measurements, it is unlikely that many industries will install laboratories for assessing the thermal properties of protective garments worn by their workers in the near future. The ideal solution would be if the manufacturers of protective garments would measure the properties and provide the values on the garments' labels.

V. HUMAN EXPOSURE ASSESSMENT

Presently, only limited data are available for objective evaluation of the toxic potential of skin contact. This is especially critical since toxic potential should be the major criterion for selection of protective garments or gloves. Dermal toxicity assessments have been constrained by limited data on the permeability of most chemicals through intact skin in vivo, the systemic toxicity of chemicals permeating the skin, and the difficulty in applying laboratory data generated through the ASTM permeation method to the human model. A few skin exposure studies have been conducted for some organic solvents such as benzene [62] and selected pesticides [8]. However, animal studies on dermal absorption of chemicals as well as rate-controlling studies of dermal drug delivery are expected to improve our level of present knowledge [6].

The exposure period calculation for most work situations should be fairly straightforward; however, coupled with a breakthrough time and steady-state permeation rate, an approximation of dermal dose could be calculated by [6]

$$\text{dose} = (T_c - T_b)(JA_e)$$

where

dose = the skin dose, mg,
T_c = the chemical contact period, sec
T_b = the time to breakthrough, sec
J = the steady-state flux of the permeating chemical, $\text{mg/m}^2 \cdot \text{sec}$
A_e = the area exposed, m^2

However, the accuracy and value of this dose estimation would be limited by a number of factors such as the influence of chemical mixtures, the quality of manufacture of the protective garment, the probability of physical failure of the barrier, individual work practices of the employee that could lead to contamination, improper or inadequate decontamination techniques, and the difficulty of applying a dermal dose estimation to the risk of systemic toxicity.

Human exposure can be monitored by passive dosimetry and biological monitoring. Human exposure to pesticides is discussed here as an example. Under the aegis of the Federal Insecticide, Fungicide, and Rodenticide Act (FIFRA), the Office of Pesticide Programs (OPP) of the Environmental Protection Agency (EPA) registers chemicals for use as pesticides provided that, among other criteria, "when used in accordance with widespread and commonly recognized practice, [they] will not generally cause unreasonable adverse effects on the environment." Protection of agricultural workers involved in the application of pesticides is considered under the umbrella of this risk criterion, and the estimation of human exposure is an integral part of the OPP's risk assessment procedures.

In 1986, EPA issued draft guidelines designed to aid pesticide registrants and others in designing and carrying out field studies that measure potential dermal and respiratory occupational exposure to pesticides, using personal monitoring techniques. These draft guidelines underwent a review by the FIFRA Scientific Advisory Panel (SAP) as well as public comment period [63].

There are two basic approaches to estimating occupational exposure to pesticides. Passive dosimetry estimates the amount of the chemical contacting the surface of the skin or the amount of the chemical available for inhalation through the use of appropriate trapping devices. Biological monitoring estimates internal dose from either a measurement of body burden in selected tissues or fluids or from the amount of pesticide or its metabolites eliminated from the body. There are theoretical and practical advantages and disadvantages to each approach.

A. Passive Dosimetry

Two major advantages of direct entrapment approaches are, first, the ability to differentiate exposure received during discrete work activities within a workday, such as mixing/loading versus applying the pesticide, and, second, the ability to compare the relative contributions of the dermal and respiratory exposure routes for each separate work activity. Table 7 shows that in many field applications, potential respiratory exposure does not contribute significantly to total exposure [63]. Table 8 indicates that for most applications, a high proportion of total dermal exposure occurs to unprotected hands. The

Table 7 Contribution of Respiratory Exposure to Total Potential Exposure for Pesticide Applicators

Activity	Physical form of pesticide used	Respiratory exposure as percent of total exposure
Drivers of tractors equipped with canopies during air-blast spraying of citrus	Spray	0.029
Drivers of ordinary tractors during boom spraying of tomatoes	Spray	0.04
Applicators using handgun to spray aquatic weeds from airboat	Spray	ND[a]
Applicators using aerosol generator for mosquito control	Aerosol	9.1
Applicators using hand knapsack mister for spraying tomatoes	Mist	3.1

[a]None detectable.
Source: Adapted from Ref. 63.

advantages of passive dosimetry are extremely important for evaluating exposure reduction safety practices such as personal protective clothing and equipment for the various work activities associated with pesticide application. For example, if a substantial fraction of total exposure for a workday occurs to the hands, forearms, and face during the short period of time a worker pours a concentrated pesticide formulation into a mix tank, risk may be mitigated significantly by requiring chemical-resistant gloves, a face shield, a closed mixing system, or some combination of these, for this short-duration/high-exposure period [63].

Traditionally, applicator exposure monitoring studies conducted for EPA have employed passive dosimetry techniques. Thus, large generic data bases have been generated from studies conducted with different pesticides [64].

Another advantage in using passive dosimetry is that the study participants are typically under the supervision of the investigator during the entire period when exposure data are being collected. The supervision ensures that good quality assurance practices are conducted and that sample integrity is not compromised [63].

A major disadvantage of passive dosimetry is that these techniques measure only the amount of chemical potentially available for absorption; independent estimates of dermal and lung absorption are required to estimate

Table 8 Contribution of Hand Exposure to Total Potential Dermal Exposure for Pesticide Applicators

Activity	Hand exposure as percent of total dermal exposure
Drivers of tractors equipped with canopies during air-blast spraying of citrus	41
Drivers of ordinary tractors during boom spraying of tomatoes	25
Applicators using handguns to spray aquatic weeds from airboats	47
Aerial applicators	55
Applicators spraying lawns, trees, and gardens with power sprayers	37
Applicators spraying lawns with hose-end sprayers	98

Source: Adapted from Ref. 63.

dosage for hazard assessment purposes. In addition, when passive dosimetry is used, it is difficult to assess the value of the clothing worn on the interception of pesticide residues.

Another potential source of error in the use of patches for residue collection is the extrapolation from the residues on the relatively small surface area of the trapping devices to entire body surface areas. Also, when the patch technique is used, crucial areas of exposure may be missed, depending on the location of the trapping devices [65].

B. Biological Monitoring

Under proper conditions, actual internal dose may be estimated from the results of biological monitoring studies. The use of information derived from biological monitoring studies varies from the early detection of a health effect to the establishment of a core relation between concentration of a chemical in fluids to absorbed dose.

For pesticides, urinary metabolites have been used to detect exposure during field operations. Swan [66] measured paraquat in the urine of applicators; Gollop and Glass [67] and Wagner and Weswig [68] measured arsenic in timber applicators; Lieben et al. [69] and Durham and Wolfe [70] measured paranitrophenol in urine after parathion exposure. A chlorobenzilate metabolite was detected in citrus workers by Levy et al. [71]; phenoxy acid herbicide metabolitites, in farmers by Kolmodin-Hedman et al. [72]; and organophosphate metabolites, in the urine of people exposed to mosquito treatments by Kutz and Strassman [73]. Davies et al. [74] used urinary metabolites of organophosphate and carbamate to confirm poisoning cases.

Even though biological monitoring offers a distinct advantage, there are inherent difficulties in study design, execution, and data interpretation, The experimenter must have a broad knowledge of the chemical being tested, and the pharmacokinetics of the chemical of interest must be known and fully understood so that the appropriate tissue, fluid, or excretion pathway, as well as the appropriate time period for monitoring, can be chosen with regard to this information [63]. Because this level of detailed information is currently unavailable for many pesticides or other chemicals, extrapolations back to the actual dose, essential in the risk assessment process, are difficult.

Biological monitoring also lacks the definition of source of exposure provided by the passive dosimetry method, but results of an appropriate biological monitoring study provide an integrated exposure picture. However, biological monitoring should be considered a chemical specific approach.

VI. DECONTAMINATION AND REUSE

A key element in the cost/benefit equation is the potential for decontamination and reuse of garments or gloves. Effective decontamination of reusable protective clothing can be critical, as shown in a study by the Vermont Health Department and NIOSH. The study indicated elevated mercury levels in the children of workers employed in a mercury thermometer manufacturing plant. These workers were wearing their work clothing home for laundering [6].

Single-use items are cost effective for some low-risk or nonroutine jobs, but for routine exposures or highly toxic exposure requiring expensive, fully encapsulating suits, decontamination and reuse may be critical. Many studies have been conducted on refurbishing pesticide-contaminated clothing (75–84), and recommendations have been made for appropriate decontamination procedures. The details of these studies are presented by Laughlin in Chapter 5 of this volume. However, much work needs to be done on nonporous elastomeric materials since establishment of a toxic reservoir in elastomeric suits and gloves is a possibility. Research performed in the area of controlled release technology has demonstrated that organic compound reservoirs can be created in elastomers that will slowly release over exceptionally long periods of time [85]. In addition, this "matrix release" could be enhanced by liquid contact from body perspiration. The assumption of potential matrix release after decontamination could be tested by evaluating protective clothing materials after several uses and decontamination in a standard ASTM permeation test cell with deionized water as the challenge liquid.

An additional factor in evaluating the effectiveness of chemical-protective clothing is the possibility of entrapment of a substance against the skin. Specifically, pinholes, poorly sealed seams, or other material imperfections could allow penetration of a hazardous liquid chemical into the clothed skin area. The protective garment would then act as a barrier to increase the gradient for movement of the liquid into the skin, thereby increasing rather than decreasing the risk. This situation could also occur through poor work practices [86] in which the garment or glove is not properly sealed against splash.

VII. CONCLUSIONS

Selection of CPC is a complex task. As discussed earlier, there are many factors that govern barrier effectiveness of a protective clothing system. These include chemical inertness of the CPC against target chemicals, permeation resistance, structural integrity, dexterity and human comfort requirements, refurbishability of reusable items, and the cost/benefit ratio.

The use of chemical protective clothing is but one component of the overall program for maintaining the health and safety of workers. It complements work practices, engineering and administrative controls, and personal hygiene. Depending upon a specific task to be performed, a worker's protective clothing may include everyday clothing with a repellent finish (fluorochemical finish) in conjunction with protective gloves and boots only; for example, for someone who is applying pesticides in lawns and gardens or for print shop workers. On the other hand, agricultural pesticide mixer/loaders/applicators may require additional protective clothing items such as repellent coveralls, face shields, and respirators. Emergency response teams, such as the coast guard or fire fighters in chemical spill situations or cleanup crews at a hazardous waste site may require fully encapsulated body suits made of impermeable plastic or elastomeric materials and equipment. It must be kept in mind that, in general, no one clothing material will be a barrier to all chemicals. Therefore, selection of CPC must be made carefully by an individual (industrial hygienist) who has a background in the physical sciences, especially chemistry, and has information about the identify of the chemicals that the CPC may be exposed to. Also, information about the degree of hazard with which the worker may be faced is of prime importance.

Much information is available regarding selection of CPC in scientific and technical literature, specific manuals [87–90], performance index books, or electronic information systems [46] as well as recommendations by the manufacturers and vendors of CPC. However, for certain chemicals or combination of chemicals, there is no commercially available glove or clothing that will provide long hours of protection following contact. In these cases, clothing must be changed as soon as it is safely possible.

Design and construction features can influence barrier effectiveness and other performance characteristics as well. For example, stitched seams of clothing may be highly permeable by chemicals if not sealed with a coating or tape; pinholes, which must be checked, may exist in elastomeric products; thickness may vary from point to point in PC items such as gloves; and garment closures depending upon style may influence chemical penetration, hence the health and safety of workers.

It is expected that sources for CPC recommendations will increase in future years because of availability of scientific literature and government regulations and standards on a worldwide basis.

REFERENCES

1. *The Prevention of Occupational Skin Disease*, 3rd printing, Soap and Detergent Association, New York, 1981.
2. *A Summary of the NIOSH Open Meeting on chemical Protective Clothing*,'' Rockville, Md., 3 June 1981, National Institute for Occupational Safety and Health, Cincinnati, Ohio, 1981.
3. S. Z. Mansdorf and B. Miles, A protective dermal film system, Paper presented at the American Industrial Hygiene Association Conference, Los Angeles, Calif., 1978.
4. Report of the advisory committee on cutaneous hazards, Report to the Assistant Secretary of Labor, Occupational Safety and Health Administration, Washington, D.C., 19 December 1978.
5. Occupational injuries and illness in the United States by industry, 1980 Bulletin 2130, Bureau of Labor Statistics, Washington, D.C. 1982.
6. S. Z. Mansdorf, Risk assessment of chemical exposure hazards in the use of chemical protective clothing—An overview, *Performance of Protective Clothing*, ASTM STP 900 (R. L. Barker and G. C. Coletta, eds.), American Society for Testing and Materials, Philadelphia, 1986, pp. 207–213.
7. M. Hougaard, The need for international cooperation regarding approvals, *Performance of Protective Clothing: Second Symposium*, ASTM STP 989 (S. Z. Mansdorf, R. Sager, and A. P. Nielsen, eds.), American Society for Testing and Materials, Philadelphia, 1988, pp. 7–14.
8. S. Libich, J. D. To, R. Frank, and G. J. Sirons, Occupational exposure of herbicide applicators to herbicides used along electric power transmission line right-of-way, *American Industrial Hygiene Association Journal 45*:56 (1984).

9. *Occupational Safety and Health Reporter*, Vol. 14; Bureau of National Affairs, Washington, D.C. 1984, p. 145.
10. R. A. Jamke, Understanding and using chemical permeation data in the selection of chemical protective clothing, *Chemical Protective Clothing Performance in Chemical Emergency Response*, ASTM STP 1037 (J. L. Perkins and J. O. Stull, eds.), American society for Testing and Materials, Philadelphia, 1989, p. 11.
11. J. O. Stull, Development of a U.S. Coast Guard chemical response suit. Report CG-D-16-87, U.S. Coast Guard, Washington, D.C. 1987.
12. R. G. Arnold, A. L. Barney, and D. C. Thompson, Fluoroelastomers, *Rubber Chemistry and Technology 46* (3):619 (1973).
13. S. P. Raikhman, V. A. Sinitsyna, V. A. Cherednichenko, and V. Shcherbakov, *Tekstil. Prom 44* (7):58 (1984).
14. E. I. du Pont de Nemours & Co., Inc., Nonwovens Rep. Int., Dec. 12, 1984.
15. E. I. du Pont de Nemours & Co., Inc., Text. Month, April 14, 1985.
16. J. F. Stampler, M. J. McLeod, M. R. Belts, A. M. Martinez, and S. P. Berardinelli, *Amer. Industr. Hyg. Assoc. J. 45*:642 (1984).
17. P. N. Mehta, *Text Res. J. 50*:185 (1980).
18. L. Benisek, in *Protective Clothing*, Shirley Publication S45, Shirley Institute, Manchester, 1982, p. 13.
19. S. K. An, R. L. Barker, and J. O. Stull, *Chemical Protective Clothing Performance in Chemical Emergency Response*, ASTM STP 1037 (J. L. Perkins and J. O. Stull, eds.), American Society for Testing and Materials, Philadelphia, 1989, p. 11.
20. A. D. Schwope, R. Goydan, D. J. Ehntholt, U. Frank, and A. P. Nielsen, Resistance of glove materials to permeation by agricultural pesticides, *Performance of Protective Clothing: Fourth Volume, ASTM STP 1133* (J. P. McBriarty and N. W. Henry, eds.), American Society for Testing and Materials, Philadelphia, 1992.
21. J. Economy and R. Y. Lin, *Applied Polymer Sciences Symposium*, No. 29, John Wiley, New York, 1976, p. 199.
22. R. Goydan, A. D. Schwope, R. C. Reid, S. Krishnamurthy, and K. Wong, Approaches to predicting the cumulative permeation of chemicals through protective clothing polymers, *Performance of Protective Clothing: Second Symposium*, ASTM STP 989 (S. Z. Mansdorf, R. Sager, and A. P. Nielsen, eds.), American Society for Testing and Materials, Philadelphia, 1988, pp. 257–268.
23. N. W. Henry, III, and C. N. Schlatter, *Am. Ind. Hygiene Assoc, J. 42*:202 (1981).
24. C. B. Billing, Jr., and A. P. Bentz, Effect of temperature, material thickness, and experimental apparatus on permeation measurement, *Performance of Protective Clothing*, ASTM STP 989 (S. Z. Mansdorf, R. Sager, and A. P. Nielsen, eds.), American Society for Testing and Materials, Philadelphia, 1988, pp. 257–268.
25. J. Crank, *The Mathematics of Diffusion*, 2nd ed., Clarendon Press, Oxford, U.K., 1975.

26. D. C. Bomberger, S. K. Brauman, and R. T. Podoll, Studies to support PMN review: Effectiveness of protective gloves, report to Environmental Protection Agency under Contract 68-[1] 6016, SRI International, Menlo Park, Calif., September 1984.
27. A. D. Schwope, R. Goydan, and R. C. Reid, Breakthrough time—What is it?, paper presented at the 2nd Scandinavian Symposium on Protective Clothing Against Chemicals and Other Health Risks, Stockholm, 5–7, November 1986.
28. G. C. Coletta, Development of performance criteria for protective clothing used against carcinogenic liquids, report to National Institue for Occupational Safety and Health Under Contract 210-76-0130, Arthur D. Little, Cambridge, Mass., October 1978.
29. K. L. Verschoor, L. N. Britton, and E. D. Golla, Innovative method for determining minimum detectable limits in permeation testing, *Performance of Protective Clothing: Second Symposium*, ASTM STP 989 (S. Z. Mansdorf, R. Sager, and A. P. Nielsen, eds.), American Society for Testing and Materials, Philadelphia, 1988, pp. 252–256.
30. M. Ridge and J. L. Perkins, Permeation of solvent mixtures through protective clothing elastomers, *Chemical Protective Clothing Performance in Chemical Emergency Response*, ASTM STP 1037 (J. L. Perkins, and J. O. Stull, eds.), American Society for Testing and Materials, Philadelphia, 1989, p. 115.
31. R. M. Barrer, *Diffusion In and Through Solids*, Cambridge at the University Press, 1941 (1st printing), 1951 (reprinted with corrections).
32. F. W. Billmeyer, Jr., *Textbook of Polymer Science, 3rd Ed., John Wiley & Sons, New York, 1984.*
33. *C. M. Hansen, The universality of the solubility parameter, I&EC Prod. Res. and Dev. 8* (1):2 (1969).
34. J. D. Crowley, G. S. Teague Jr., and J. W. Lowe, A 3D approach to solubility, *J. Paint Technology 38* (496):269 (1966).
35. R. E. Morris and P. T. Wagner, Swelling of nitrile rubbers by iso-octane-Toluene blends, *Ind. Eng. Chem. 49* (3):445 (1957).
36. A. I. Abu-Isa, Elastomer–gasoline blends interactions. I. Effects of methanol–gasoline mixtures on elastomers, *Rubber Chemistry and Technology 56* (1):135 (1983).
37. A. I. Abu-Isa, "Elastomer–gasoline blends interactions. II. Effects of ethanol/gasoline and methyl-t-butyl ether/gasoline mixtures on elastomers, *Rubber Chemistry and Technology 56* (1):169 (1983).
38. M. E. Myers and I. A. Abu-Isa, Elastomer–gasoline blends interactions. III—Effects of methanol mixtures on fluorocarbon elastomers, *Journal of Applied Polymer Science 32* (2):3515 (1986).
39. G. J. van Amerongen, Diffusion in elastomers, *Rubber Chem. Technol. 37*:1065 (1964).
40. A. S. Michaels and H. J. Bixler, Membrane permeation: theory and practice, *Progress in Separation and Purification* (E. S. Perry, ed.), Wiley Interscience, New York, 1968.

41. R. D. Bennett, C. E. Feigley, E. O. Oswald, and R. H. Hill, The permeation by liquefied coal of gloves used in coal liquefaction pilot plants, *Am. Ind. Hyg. Assoc. J. 44* (6):447 (1983).
42. S. L. Davis, C. E. Feigley, and G. A. Dwiggins, *A Comparison of Two Methods Used to Measure Permation of Glove Material by a Complex Organic Mixture*, School of Public Health, University of South Carolina.
43. K. Forsberg and F. Stathoula, The permeation of multi-component liquids through new and pre-exposed glove materials, *Am. Ind. Hyg. Assoc. J. 47* (3):189 (1986).
44. R. L. Mickelsen, M. M. Roder, and S. P. Berardinelli, Permeation of chemical protective clothing by three binary solvent mixtures, *Am. Ind. Hyg. Assoc. J. 47* (4):236 (1986).
45. A. D. Schwope, P. P. Costas, J. O. Jackson, and D. J. Weitzman, *Guidelines for the Selection of Chemical Protective Clothing*, Vols. I and II, Arthur D. Little, Cambridge, MA, 1983.
46. K. Forsberg and L. H. Keith, *Chemical Protective Clothing Performance Index*, Wiley-Interscience, New York, 1989.
47. A. M. Barton, Solubility parameters, *Chemical Reviews 75* (6): (1975).
48. L. Snyder, Solutions to solution problems I, *Chemtech* (1979).
49. L. Snyder, Solutions to solution problems II, *Chemtech* (1980).
50. M. Raheel and E. C. Gitz, Effect of fabric geometry on resistance to pesticide penetration and degradation, *Arch Environ. Contam. Toxicol. 14*:273 (1985).
51. M. Raheel, Resistance of selected textiles to pesticide penetration and degradation, *J. Environ. Health 49* (4):214 (1987).
52. M. Raheel, Barrier effectiveness of apparel fabrics toward pesticide penetration, *J. Environ. Health 51* (2):82 (1988).
53. M. Raheel, Pesticide penetration in fabrics: Fiber chemistry, surface energy, and fabric porosity, *Technical Papers, the First International Symposium on the Impact of Pesticides, Industrial and Consumer Chemicals on the Near Environment* (Reagan, Johnson, and Dusaj, eds.), 1988, pp. 127–136.
54. D. A. Branson, G. Ayers, and M. Henry, ''Effectiveness of selected work fabrics as barriers to pesticide penetration.'' *Performance of protective clothing*, ASTM STP 900 (Barker and Coletta, eds.), 1986, pp. 114–120.
55. M. Raheel, Pesticide transmission in fabrics: Effect of perspiration, *Bull. Environ. Contam. Toxicol. 46*:837 (1991).
56. M. Raheel, Pesticide transmission in fabrics: Effect of particulate soil, *Bull. Environ. Contam. Toxicol. 46*:845 (1991).
57. R. Goydan, A. D. Schwope, and R. C. Reid, Methodologies for estimating protective clothing performance—Development, validation and assessment, Interim Draft Report of EPA Contract 68-030-3293, Work Assignment 08, to Arthur D. Little, Inc., 1986.
58. J. Lara, N. F. Nelisse, S. Cote, and H. J. Nelisse, Development of a method to evaluate the puncture resistance of protective clothing materials, *Performance of Protective Clothing: Fourth volume*, ASTM STP 1133 (J. P. McBriarty and N.

W. Henry, eds.), American Society for Testing and Materials, Philadelphia, 1992.

59. V. H. Freed, T. E. Davies, L. J. Peters, and F. Parveen, *Residue Reviews* *75*:159 (1980).
60. I. Holmer, Thermal properties of protective clothing and prediction of physiological strain, *Performance of Protective Clothing: Second Symposium*, ASTM STP 989 (S. Z. Mansdorf, R. Sager, and A. P. Nielsen, eds.), American Society for Testing and Materials, Philadelphia, 1988, pp. 82–86.
61. B. W. Olesen and F. N. Dukes-Dobos, International standards for assessing the effect of clothing on heat tolerance and comfort, *Performance of Protective Clothing: Second Symposium*, ASTM STP 989 (S. Z. Mansdorf, R. Sager, and A. P. Nielsen, eds.), American Society for Testing and Materials, Philadelphia, 1988, pp. 17–30.
62. G. O. Nelson, B. U. Lum, G. J. Carlson, C. M. Wong, and J. S. Johnson, Glove permeation by organic solvents, *Am. Ind. Hygiene J. 42*:217 (1981).
63. C. Lunchick, A. P. Nielsen, J. C. Reinert, and D. M. Mazzetta, Pesticide applicator exposure monitoring: EPA guidelines, *Performance of Protective Clothing: Second Symposium*, ASTM STP 989 (S. Z. Mansdorf, R. Sager, and A. P. Nielsen, eds.), American Society for Testing and Materials, Philadelphia, 1988, pp. 756–771.
64. D. C. Eberhart, E. W. Day, R. D. Knarr, D. M. Zimmerman, A. P. Nielsen, J. C. Reinert, and N. I. Muir, Progress in the development of a generic data base for mixer/loader—applicator exposures, *Performance of Protective Clothing: Second Symposium*, ASTM STP 989 (S. Z. Mansdorf, R. Sager, and A. P. Nielsen, eds.), American Society for Testing and Materials, Philadelphia, 1988, pp. 772–775.
65. C. A. Franklin, R. A. Fenske, R. Greenhalgh, L. Mathieu, H. V. Denley, J. T. Leffingwell, and R. C. Spear, *Journal of Toxicology and Environmental Health, 7* (5):715 (1981).
66. A. A. B. Swan, *British Journal of Industrial Medicine, 26*:322 (1969).
67. B. R. Gollop and W. I. Glass, *New Zealand Medical Journal, 89*:10 (1979).
68. S. L. Wagner and P. Weswig, *Archives of Environmental Health, 28*:77 (1974).
69. J. R. Lieben, K. Waldman, and L. Krause, *Industrial Hygiene and Occupational Medicine, 7*:93 (1953).
70. W. F. Durham and H. R. Wolfe, *Bulletin of the World Health Organization, 26*:75 (1962).
71. K. A. Levy, S. S. Brady, and P. Pfaffenberger, *Bulletin of Environmental Contamination and Toxicology 27* (2):235 1081.
72. B. Kolmodin-Hedman, S. Hoglund, and M. Okerblom, *Archives of Toxicology and Contamination 54*:267 (1983).
73. F. W. Kutz and S. C. Strassman, *Mosquito News 37* (12):211 (1981).
74. J. E. Davies, H. F. Enos, A. Barquet, C. Morgade, and J. X. Danauskas, in *Toxicology and Occupational Medicine* (W. B. Deichman, ed.), Elsevier, New York, 1979, pp. 369–380.

75. C. B. Easley, J. M. Laughlin, R. E. Gold, and R. M. Hill, Laundry factors influencing methyl parathion removal from contaminated denim fabric, *Bulletin of Environmental Contamination and Toxicology 29* (4):461 (1982).
76. C. B. Easley, J. M. Laughlin, R. E. Gold, and D. Tupy, Laundering procedures for removal of 2, 4-dichlorophenoxyacetic acid ester and amine herbicides from contaminated fabrics, *Archives of Environmental Contamination and Toxicology 12* (1):71 (1983).
77. C. B. Easley, J. M. Laughlin, R. E. Gold, and D. Tupy, Methyl parathion removal from work weight fabrics by selected laundry procedures, *Bulletin of Environmental Contamination and Toxicology 27* (2):101 (1981).
78. J. L. Keaschall, J. M. Laughlin, and R. E. Gold, Effect of laundering procedures and functional finishes on removal of insecticides selected from three chemical classes, *Performance of Protective Clothing*, ASTM STP 900 (R. L. Barker and G. C. Coletta, eds.), American Society for Testing and Materials, Philadelphia, 1986, pp. 162–176.
79. J. M. Laughlin, C. B. Easley, R. E. Gold, and R. M. Hill, Fabric parameters and pesticide characteristics that impact on dermal exposure of applicators, *Performance of Protective Clothing*, ASTM STP 900 (R. L. Barker and G. C. Coletta, eds.), American Society for Testing and Materials, Philadelphia, 1986, pp. 136–150.
80. T. H. Lillie, J. M. Livingston, and M. A. Hamilton, Recommendations for selecting and decontaminating pesticide applicator clothing, *Bulletin of Environmental Contamination and Toxicology 27* (6):716 (1981).
81. J. M. Laughlin and R. E. Gold, Removal and redeposition of methyl parathion during laundering of functionally finished textiles, Technical Papers, *The First International Symposium on the Impact of Pesticides, Industrial and Consumer Chemicals on the Near Environment* (B. Reagan, D. Johnson, and S. Dusaj, eds.), sponsored by USDA-CSRS, 1988, pp. 12–14.
82. C. J. Kim, J. F. Stone, and C. E. Sizer, *Bulletin of Environmental Contamination and Toxicology 29* (1):95 (1982).
83. C. J. Kim, J. F. Stone, J. R. Coats, and S. J. Kadolph, *Bulletin of Environmental Contamination and Toxicology 36* (2):234 (1986).
84. J. M. Laughlin, C. B. Easley, R. E. Gold, and R. M. Hill, Fabric parameters and pesticide characteristics that impact on dermal exposure of applicators, *Performance of Protective Clothing*, ASTM STP 900 (R. L. Barker and G. C. Coletta, eds.), American Society for Testing and Materials, Philadelphia, 1986, pp. 136–150.
85. S. Z. Mansdorf, Slow release: Development of aquatic herbicides and carrier systems, *Proceedings, Controlled Release Pesticides Symposium*, Controlled Release Society, University of Akron, Ohio, 1974, pp. 121–127.
86. H. N. Nigg, J. H. Stamper, and R. M. Queen, *American Industrial Hygiene Association Journal 45*:182 (1984).
87. K. Forsberg and K. G. Olsson, *Guidelines for the Selection of Chemical Protective Gloves*, The Royal Institute of Technology, Stockholm, May 1985 (in

Swedish) (available through Foreningen for Arbetarshydd, Kungsholm Hamnmplan 3, 11220 Stockholm, Sweden).

88. The Products Control Board, *Application of the Labeling Ordinance*; The National Swedish Environment Protection Board, Advice and Guidelines, March 1983 (available through Liber Distribution, Forlagsorder, 16289 Stockholm, Sweden).
89. *NIOSH/OSHA Pocket Guide to Chemical Hazards*, DHEW (NIOSH) Publication No. 78-210, National Institute of Occupational Safety and Health/Occupational Safety and Health Administration, Cincinnati, Ohio.
90. Edmont Gloves and Protective Equipment, *Edmont Chemical Resistance Guide*, Edmont Division Becton, Dickinson & Co., Coshocton, Ohio, 1983.

4

Attitudes and Clothing Practices of Pesticide Applicators

MARGARET RUCKER University of California at Davis, Davis, California

I. INTRODUCTION

Statistics on pesticide usage suggest that applicator exposure is an important public health problem. It has been estimated that by the late 1980s anywhere from 4 to 7 billion pounds of agricultural pesticides were being used worldwide [1,2]. A recent estimate in the Wall Street Journal [3] is that, in the United States alone, synthetic insecticide sales are over $1 billion per year.

Initial concerns about pesticide exposure focused on short-term effects. It was recognized that not only were high doses lethal but low doses also could have a variety of ill effects. For example, Namba et al. [4] reported mental confusion and drowsiness as symptoms of acute organophosphate poisoning. Kahn's [5] summary of short-term symptoms resulting from pesticide exposure included nausea, eye and skin irritation, headaches, and insomnia. A number of long-term effects have also been reported. For instance, work by Smith, Stavinoha, and Ryan [6] as well as Metcalf and Holmes [7] related chronic organophosphate exposure to difficulties in focusing attention and maintaining alertness. Work by Savage et al. [8] suggests that chronic neurological problems may follow acute organophosphate poisonings. A number of studies have indicated that cancer may be another long-term effect of pesticide use [9–14].

Figures for number of deaths per year in the United States that are attributable to pesticides range from 50 to 250 [15]. Estimates of the number of people experiencing some symptoms of pesticide poisoning also vary with the reporting system. For example, the average annual number of occupational cases in California in the 1980s was set at about 1,100, and the average annual number of pesticide illnesses among field workers was reported to be 345. However, these figures did not include data on exposures to antimicrobials; inclusion of these data added more than 700 additional cases [16].

Furthermore, the poisonings that do get entered into formal reporting systems are apt to be only a small portion of those that actually occur. In their survey of farm laborers in Tulare County, California, Mines and Kearney [17] found that the majority of workers who experienced ill effects from use of chemicals on the job did not report the problem to their employer. When asked whether they contacted a doctor about the health problem, over three-fourths of the workers said they did not seek medical attention from a physician. The main reasons given for this failure to report included concern about the cost of medical care and unwillingness to miss work. Other factors related to agricultural workers' failure to report included fear of being fired or deported if they called attention to themselves. A subsequent study [18] of agricultural workers in Ventura County, California, indicated that in addition to these causes of underreporting, workers may not be aware that they should report a job-related illness or injury or may not know how to file for benefits.

Attempts to reduce applicators' exposure to agricultural chemicals have taken several directions. One of these is to develop pest-resistant varieties of common crops. Efforts have been successful in making a few plants resistant to some pests, but much more work needs to be done. Other efforts have focused on closed systems for the pesticides and closed environments for the applicators. Again, these efforts have shown some success, but the goal of eliminating exposure through better systems engineering has proved to be elusive. A third approach has been to improve the garments available for pesticide applicators as well as the attitudes and practices with respect to the use of these garments. Advances in garment development have been discussed elsewhere; data on applicator attitudes and practices will be reviewed in this chapter.

II. PESTICIDE APPLICATORS' CLOTHING PRACTICES

Articles as far back as the 1950s provide anecdotal evidence for problematic practices by at least some applicators. For example, in a 1951 article on parathion hazards, Griffiths, Stearns, and Thompson [19] made mention of one worker experiencing a marked cholinesterase drop due to drift exposure to the point that his skin and clothing were saturated. The authors also reported finding that recommendations to bathe and change clothing at the end of each day were not always followed by their sample of applicators. Quinby and Lemmon [20] attributed several of the pesticide poisoning cases in their study of parathion exposure to not wearing protective clothing in some instances and continued wearing of contaminated clothing in others. Another potentially hazardous practice is noted in a report by Culver, Caplan, and Batchelor [21]. The authors describe the case of a mixer-loader who wore long rubber gauntlets throughout the time he was working and still managed to register hand exposure to the pesticide in the range of 27 to 80 mg. The investigators' assumption was that this high level of exposure was due to inappropriate techniques in putting on the gloves and removing them.

A subsequent article by Wolfe [22] based on his experiences at the Wenatchee Research Station in Wenatchee, Washington, acknowledged that glove contamination almost always occurs to some degree. However, it was his contention that unless there was "gross" contamination of the glove interior, pesticide applicators are at less risk when they wear gloves than when they do not. Wolfe was less definite, however, about the advisability of covering the body with other articles of protective clothing such as waterproof jackets once the skin had been contaminated. He observed that this often happened in practice because applicators tended to wait until exposure from pesticide drift was obvious before donning protective clothing.

A report by Rogers [23] on attitudes and practices of pesticide applicators in Mississippi, based on 20 years of interactions with tractor drivers, aerial applicators, farm managers, and other consultants, suggests that practices may differ by region. He remarked on the rare wearing of gloves in the Mississippi Delta; he attributed the common occurrence of this unsafe practice to use of highboy equipment and consequent time pressures on the applicators, but it is likely that the warm climate limited glove use as well. Rogers also noted that the growers, quite reasonably, adjusted their practices according to the chemicals they were using. That is, early in the cotton season there was less evidence of precautionary practices than late in the season because less toxic pesticides were applied early in the season.

In response to recognition of a need for more than anecdotal data on which to base pesticide safety programs, Beal et al. [24] interviewed 229 Iowa farmers about their pesticide use and expenditure patterns. In this survey, however, the focus was on *what* was used rather than *how* it was used. Even so, their data on usage did serve to confirm the extent and variety of chemicals applied for agricultural pest control.

A series of studies reported in the 1980s provided more detailed information about clothing practices of pesticide applicators and also family members who took care of the applicators' clothing. This series includes work by

DeJonge et al. [25], Keeble et al. [26], Perkins et al. [27], and a regional project involving a five-state comparison of applicator practices [28] as well as launderer practices [29]. Additional information from states participating in the regional project also appeared in Rucker et al. [30], Stone et al. [31], and Stone et al. [32].

DeJonge et al. [25] investigated the relationship between demographic data and applicator practices in a sample of 469 private licensed applicators in Michigan. In a comparison of farmers growing fruit, cultivating field or vegetable crops, and raising livestock, fruit growers were found to take protective action significantly more often than the other two types of farmers. Fruit growers also believed more in the importance of taking protective action. These findings were taken as support for the theory of Rogers and Shoemaker [33] that problem recognition is a critical initial step in the adoption process. However, subsequent work by Rucker et al. [30] suggested that in some instances knowledge of being at high risk may actually be associated with *less* interest in taking precautions. Possible explanations and implications of this relationship will be discussed further in the section on pesticide applicators' attitudes.

In the applicator survey described by Rucker et al. [28], one objective was to determine what might be common trends in practices across states versus state-specific practices. Therefore, their sample included 130 cotton growers from California, 125 cotton growers and 205 wheat growers in Oklahoma, 294 corn growers in Minnesota, 150 corn growers and 72 apple growers in Michigan, and 638 pesticide applicators from Iowa. Practices covered in the applicator survey included selection, use, and storage of the clothing worn when applying pesticides.

The data on what growers selected to wear for pesticide application were quite similar across the five states (see Table 1). A majority of respondents from each state said they usually wore long-sleeved shirts and jeans or work pants when applying pesticides. Such choices are consistent with generally accepted recommendations to cover arms and legs as well as the torso. A majority of the growers in each state also reported wearing gloves when applying pesticides; there was a range from 30 to 49% across states of growers who said their gloves were made of waterproof vinyl or rubber. However, another 20–48% reported using leather or canvas gloves. These materials are generally not recommended because they tend to absorb and retain pesticides next to the skin. In terms of foot and head coverings, the most popular items were leather shoes and company or baseball caps. Again, these items are problematic in terms of potential for absorbing and retaining pesticides [34]. Few respondents in any of the states reported wearing waterproof jackets, pants, or boots.

Table 1 Five-State Comparison of Type of Clothing Usually Worn When Applying Pesticides

Item	Response by state, %[a]					Total, %
	California	Iowa	Michigan	Minnesota	Oklahoma	
Work or sport shirts						
long sleeves	55	68	68	70	66	67
short sleeves	30	25	22	26	24	25
sleeveless	1	2	2	2	1	2
do not wear	—	—	—	—	—	—
Pants						
coveralls	31	19	26	21	16	21
bib overalls	2	12	11	9	25	13
jeans, work pants	58	69	64	73	57	66
sweat pants	—	—	—	—	—	—
shorts, cut offs	1	—	1	—	—	—
Work shoes or boots						
vinyl, rubber	17	7	12	14	11	10
leather	68	82	80	84	78	80
canvas	2	1	2	1	1	1
other	—	—	—	—	—	—
Gloves						
vinyl, rubber	49	34	37	30	42	37
leather	15	24	11	18	24	20
canvas	5	18	14	30	11	17
other	—	5	4	3	—	4
do not wear	16	11	26	16	15	15
Hats						
hard plastic	7	1	5	1	5	3
felt	4	2	3	2	6	3
straw	8	1	4	2	8	3
company, baseball	55	76	64	76	67	71
other	3	2	7	5	—	3
do not wear	8	8	8	6	2	6
Other clothes						
jacket, coat	33	42	43	52	32	41
sweatshirt	5	36	16	26	3	22
sleeveless vest	3	8	6	8	2	6
undershirt	34	45	42	48	41	43
jockey, boxer shorts	48	60	50	60	53	56
socks	55	68	57	67	64	64
belt	44	56	47	57	45	52
waterproof jacket	8	1	13	2	5	4
waterproof pants	5	1	8	1	3	3
other	3	1	3	1	2	1

[a]Percentages in a given category may total more or less than 100% due to nonresponses or multiple responses.
Source: Adapted with permission from Ref. 28.

In terms of reuse of pesticide-soiled clothing from one day to the next, this was not generally a common practice. However, responses varied considerably from state to state, from lows of 4 and 7% in Oklahoma and California to a high of 26% in Minnesota. Moreover, for those growers who did reuse clothing, most reported that it was worn again for 2 days or more. This type of reuse is not recommended since field studies indicate it can increase the risk of secondary exposure to pesticides and produce symptoms of pesticide poisoning [35,36].

Reports of changing clothes immediately if exposed to a full-strength liquid concentrate spill or if saturated with spray varied quite a bit across states. Reports of taking immediate action ranged from 44 to 71% in response to a spill and from 52 to 74% in response to saturation with spray.

Compliance with recommendations to store applicators' garments separately from other household laundry varied with the type of garment. That is, the majority of applicators reported storing outer garments separately but underwear with other laundry. Figure 1 shows the differences by garment type in each of the five states. Similar values were found in the parallel survey of people in the household who were responsible for care and maintenance of the clothing [29]. These responses suggest the need for more awareness of the fact that, as noted by Laughlin et al. [37], any article of clothing has the potential for being a source of secondary contamination.

Further analysis of the launderer data provided by Nelson et al. [29] indicated that, with the exception of washing pesticide-contaminated clothing separately and using hot or warm water, many farm families had not adopted recommended procedures for the care of pesticide-contaminated clothing. For example, slightly less than half of the launderers always used a clothesline rather than a dryer for the pesticide wash. The benefits of line drying include that the dryer does not get contaminated and, assuming the line is located outdoors, sunlight may help degrade any residue remaining after the wash [38,39]. As for reducing contamination in the washer by running it through an extra cycle with detergent, only 28% of the launderers reported taking this precaution.

In the study conducted by Keeble et al. [26], 175 fruit growers and workers from Virginia, West Virginia, and North Carolina were asked to identify what they would wear when mixing and applying two different types of pesticides. These hypothetical data are consistent with what growers in the Rucker et al. [28] survey said they actually were wearing. That is, the majority said that they would wear work shirts and pants when mixing and spraying Parathion or Captan. Even for the more toxic Category 1 pesticide, only about 25% of the sample said they would wear water-repellent coats or waterproof boots and about 50% said they would wear waterproof gloves.

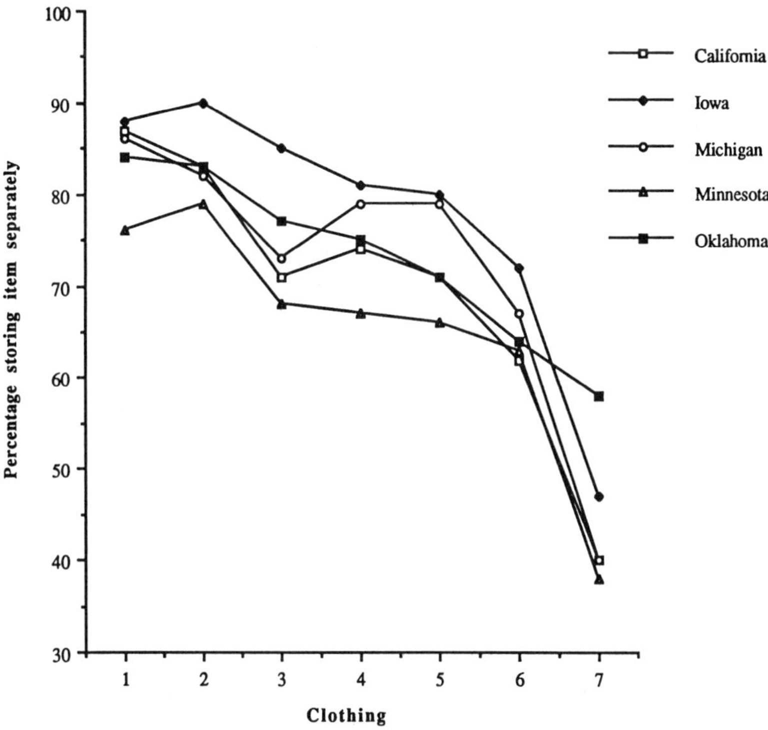

Figure 1. Five-state comparison of applicator reports of storing application clothing separate from other laundry. (From Ref. 28—Table 9).

Other recent observations and surveys with growers in the United States and Canada support the findings of the multistate surveys already described. That is, even though most growers take some precautions when applying pesticides, there is still room for improvement. For example, Helmers et al. [40] surveyed 28 agribusiness workers and 71 farmers in Iowa concerning precautions taken when distributing or applying organophosphates. The respondents in this study "utilized little more than simple hygiene measures for their own protection." That is, almost everyone reported bathing and changing clothes as protective activities. However, only about half wore rubber gloves and very few used goggles or respirators. The majority of the farmers wore a hat or cap and a long-sleeved shirt; less than half of the agribusiness workers reported doing the same. The authors proposed that the lower rate of use of protective clothing among the agribusiness workers led to their showing a statistically significant cholinesterase reduction after exposure, whereas the farmers did not.

Research in Canada suggests that Canadian applicators' protective clothing practices are not much different from those of U.S. applicators. For example, in a study of seven prairie grain farmers, Hussain et al. [41] found that only one person used coveralls, gloves, and a respirator on a regular basis when handling pesticides; the rest wore ordinary work clothes only. More extensive surveys also indicate that the majority of Canadian agricultural workers do not wear special protective clothing when working with pesticides [42,43].

Although reported less frequently than the problems described previously, two other types of practices appear to be prevalent enough to deserve some mention. One is the handling of contaminated clothing immediately after pesticide application in such a way that the clothing is itself a source of contamination. The other involves engaging in activities that raise exposure levels beyond the limits within which the protective clothing was designed to protect.

One way in which hand contamination can occur following pesticide application is through touching the outside surface of the glove with the bare hand in the process of removing the gloves. This possibility has been discussed earlier. In addition, Fenske et al. [44] found that workers in their study experienced hand contamination as a result of touching their work clothing after taking off their protective gloves. These authors suggest that removal of work clothing with one's bare hands after applying pesticide may be a major source of hand exposure.

During interviews with growers about their glove preferences and problems, Rucker et al. [45] observed another unsafe practice during demonstrations of the glove removal process. That is, the typical method for adjusting inverted glove fingers was to place the mouth over the open end of the glove and exhale into the interior, creating the potential for oral contact with pesticides.

As for what may be considered undue reliance on the barrier properties of protective clothing, Fenske et al. [44] noted that pesticide-treated grain is often stirred with the hand and arm rather than by mechanical means. In their own study, they found one worker scooping the grain with his gloved hands instead of using a mixing stick.

To summarize, there has been and continues to be evidence of a variety of unsafe practices among pesticide applicators. Efforts to reduce the incidence of unsafe use of pesticides and inappropriate use of protective clothing have focused on two types of variables that could be useful in understanding and modifying behavior (i.e., knowledge and attitudes).

III. PESTICIDE APPLICATORS' KNOWLEDGE

As noted by Beal et al. [24] early attempts to ensure the safe and proper use of agricultural chemicals placed heavy emphasis on testing farmers' knowledge and developing educational programs to supplement that knowledge. However, even the proponents of a knowledge-based approach to increasing applicator health and safety admit there are some problems with this approach. For example, adequate tests are difficult to construct due to the large number of chemicals on the market and the wide variation in number and kind of chemicals used by individual farmers. There is also the question of what kind of general working knowledge farmers should be able to draw on at all times versus specific information that could be readily acquired at the time of application, from sources such as the product label.

In spite of some of the problems with knowledge-based programs, government mandated and private applicator training programs continue to flourish. Being informed does not guarantee safe practices, but being uninformed is almost certain to produce unsafe practices. As Wasserstrom and Wiles [2] note in their review of two applicator training program assessments, farmers reported a greater awareness of safety issues after training and in *some* cases modified their behavior to better protect themselves. For example, in an evaluation of certified applicator training in North Carolina, the percentage of farmers who reported wearing waterproof gloves during mixing and loading increased from 40 to 44% with training. An evaluation of training in New York, Georgia, Pennsylvania, Nebraska, and Iowa showed a much larger increase in compliance with glove regulations, from 42 to 77%. Even in the second case, although a change of this magnitude is gratifying, it means that half of the applicators who did not use waterproof gloves before the training were still not using the gloves after the training. Furthermore, there was almost no increase in willingness to wear other forms of protective clothing.

IV. PESTICIDE APPLICATORS' ATTITUDES

In an effort to understand and modify applicators' practices beyond what appears to be possible in knowledge-based programs, researchers have turned to an examination of applicators' attitudes. They have considered a variety of attitude categories related to pesticides including the importance of chemicals in agricultural productivity, as well as their potential danger to humans and potential danger to the environment.

From the early work of Beal et al. [24] up to the present time, growers generally have reported positive attitudes toward the contributions of chemicals to agriculture and have minimized potential risks to humans and the environment. In the Beal et al. study, for example, most growers agreed with statements of the type, "Agricultural chemicals are one of the primary factors contributing to the U.S. standard of living and way of life" and "The use of agricultural chemicals is a profitable input in the farmer's operation" as well as "Almost all deadly effects of agricultural chemicals are due to improper handling, a disregard for safety precautions or improper use."

Vlek and Stallen [46] found that when risk–benefit perceptions of people from different occupations were compared, agricultural personnel tended to rate pesticide application as a relatively acceptable risk. This finding was attributed to the agricultural workers' focus on the large-scale benefits with less attention paid to the potential risks. On the other hand, respondents from the medical, social, and scientific professions considered activities with large-scale benefits but also substantial risks as being relatively unacceptable. Various explanations proposed for agricultural workers finding risky activities such as pesticide application more acceptable include self-selection into the area because of greater interest and involvement, greater experience with and subsequent adaptation to the consequences, and attempts to reduce the cognitive dissonance involved in continuing to participate in activities that could be hazardous to one's health. Whatever the explanations, the difference in attitudes of growers versus people in the scientific, social, and medical professions suggests that the latter should be very careful in designing pesticide safety programs for the former. Without appreciation for the other group's perspectives, training materials are apt to miss the mark.

Although attitudes toward the overall risk involved in the use of pesticides is important to assess, it is even more important to determine how applicators view the risks associated with different types of exposure (i.e., respiratory, oral, and dermal). Wolfe et al. [47] found evidence that applicators typically underestimate dermal risks. Additional information on applicators' attitudes toward dermal risk is provided in the study by Rucker et al. [28]. In their study, applicators were asked four different questions dealing with pesticides coming in contact with their skin: probability of immediate health risk, severity of immediate health risk, probability of long-term harm, and severity of long-term harm. In all five states, the majority of applicators rated both immediate and long-term health risks as not likely (percentages of state samples ranged from 66 to 86%) and, if they did occur, not serious (percentages ranged from 59 to 81%).

Table 2 Five-State Comparison of Perceived Pesticide Exposure and Effectiveness of Protective Clothing

Item	Response by state, %[a]						Total, %
	California	Iowa	Michigan	Minnesota	Oklahoma	χ^2	
Pesticide gets on clothes:							
Reporting seldom or never	54	38	34	39	50	24.72[a]	41
Pesticide gets through clothes:							
Reporting seldom or never	77	67	66	67	69	4.00	68
Perceptions that clothing is relatively effective as protection from pesticide exposure	73	64	70	64	65	5.83	66

[a] $p < .001$.
Source: Adapted from Ref. 28.

Other attributes that obviously must be considered in understanding and modifying applicators' behavior toward clothing and other personal protective equipment are attitudes toward traditional work clothes and attitudes toward special protective clothing. In the study by DeJonge et al. [25], growers were asked for their preferences concerning color, garment type, and garment features. The most popular choice was a blue shirt, pants, and hat, made of either cotton or cotton-polyester. In other words, the growers felt most positive toward a traditional chambray work shirt and jeans type of outfit.

Other studies indicate that applicators also feel that traditional work clothing provides an effective barrier to pesticides. In the Rucker et al. [30] study of growers and commercial applicators, the average rating of the effectiveness of clothing in providing protection from pesticides was somewhat above the midpoint of the scale, toward the very effective end. In the five-state study of growers [28], percentages of respondents who felt pesticide seldom or never got on their clothes ranged from 34 to 54%. When pesticide was seen as getting on clothing, however, it was commonly assumed to go no further; a majority in each state (66–77%) said that pesticide seldom or never got through their clothing to the skin. Along that same line, a majority in each state (64–73%) rated their usual clothing as an effective, or at least not ineffective, barrier to pesticides. The complete set of data on perceived exposure and protection provided by clothing is presented in Table 2.

In contrast to the favorable attitudes expressed toward traditional work clothing, attitudes toward special protective clothing have been generally negative. Wolfe [22] detailed the two major problems with early types of protective gloves, jackets, and trousers as being thermal discomfort and restriction in movement. In the Rucker et al. [45] study of protective gloves, growers felt the gloves they could choose from were of two types—thick and loose or thin and tight. The thick gloves did not provide the dexterity needed to perform some of the tasks involved in the application process. On the other hand, the thin gloves were hot; constricted the hands, which caused fatigue and pain or numbness; and were difficult to remove. Although improvements have been made in the materials available for protective clothing, there remain some problems of worker acceptability of even many of the newer materials at very high temperatures [cf. 48]. As Perkins et al. [27] have remarked, the unpleasant experiences with early forms of protective clothing made it more difficult to gain acceptance for the newer types.

In studies focused specifically on agricultural workers, there have been attempts to explore the relationships among being at risk, risk perceptions, and willingness to take precautionary actions. DeJonge et al. [25] found that fruit growers, who were at relatively high risk compared to the field crop and livestock people, exhibited a greater belief in the importance of protection and actually wore more protective clothing. As shown in Table 3, in the study by Keeble et al. [26], there also was a positive relationship between perceptions of risk and estimated use of protective clothing.

In contrast, the Rucker et al. [30] study comparing growers' responses with commercial applicators' responses suggested that those at more risk were

Table 3 Estimated Use of Personal Protective Equipment When Applying Two Different Types of Pesticides[a]

Items of clothing and equipment	Percent[b] who would wear item when using Parathion[c]	Percent[b] who would wear item when using Captan[d]
Waterproof gloves	50.9	36.0
Respirators	45.1	26.9
Long water-repellent coats	25.7	17.1
Waterproof boots	22.9	17.7
Goggles	21.1	12.6
Water-repellent pants	19.4	13.1
Long water-repellent coats with hoods	9.1	6.9
Waterproof hat	2.9	2.3
Disposable coveralls	.6	.6

[a] $N = 175$

[b] Totals exceed 100% because multiple answers were possible.

[c] Category I pesticide.

[d] Category IV pesticide.

Source: Adapted from Ref. 26.

Table 4 Significant Differences Between Commercial Applicators and Growers: Being at Risk Versus Perceived Importance of Precautions

Variable	Applicator mean	Grower mean	χ^2
Use of restricted pesticides[a]	0.56	1.60	9.36[d]
It is important to wash your hands after handling pesticides[b]	1.06	1.40	3.89[c]
It isn't safe to let children hang around when working with pesticides[b]	1.00	1.47	6.64[d]

[a]Scale values are 0 (nonrestricted), 1 (restricted with qualifications), and 2 (restricted without qualifications).
[b]Scale ranges from 1 (strongly agree) to 5 (strongly disagree).
[c]$p < 0.05$
[d]$p < 0.01$
Source: Adapted from Ref. 30.

less likely to rate certain precautions as important, including washing the hands after handling pesticides and keeping children out of the area when working with pesticides. These data are presented in Table 4. Similarly, Helmers et al. [40] found that agribusiness workers were more likely than farmers to doubt that farm chemicals are safe if used properly, but farmers were more likely to use hats or caps and long-sleeved shirts to protect their skin. Dissonance reduction needs and defensive-avoidance reactions were suggested as explanations for an inverse relationship between perceived risk and willingness to take precautions. However, at this point, it seems that additional work is needed to clarify when there is a direct relationship between perceived risk and willingness to take action and when there is an inverse relationship.

Some of the models of attitudes and attitude change offer promise in this regard. General models such as the Fishbein Model of Attitudes and the Fishbein Extended Model have already undergone some testing as conceptual frameworks in research on attitudes and behavior of applicators toward protective clothing [27]. Other models specifically designed to accommodate considerations of risk and health protective behaviors are also beginning to be tested with pesticide applicator samples [49–51].

V. FUTURE DIRECTIONS

There is a need to continue knowledge-based applicator training programs. Not only are such programs required for certification, but they are also just a good idea. New applicators need a thorough grounding in the basics, and experienced applicators can always benefit from a refresher course. These training programs are especially critical in times of changing regulations. As Pryer [52] has pointed out, farmers have a wide variety of issues to keep informed about and therefore can benefit from some guidance through the pesticide regulation legalese.

Additional consideration of attitudes is also needed in the context of attitude/behavior models. It is time to move beyond simple descriptions of attitudes and practices. We need to gain a better understanding of the linkages among beliefs, attitudes, values, and cues to action as they influence willingness to use protective clothing and engage in other precautionary behaviors. Some starting points include Fishbein's basic and extended models [53,54] and variations of the Health Belief Model [cf. 55].

Finally, we need to take a systems approach to the physical and social environment as well as the psychological environment. For instance, instructions for care of contaminated clothing that are relevant for a middle class Illinois farmer may not be relevant to a low-income migrant worker in California. Therefore, if an informational booklet on pesticide safety is to have a positive impact on a wide audience, it will often have to present options from which the different groups can choose. Even within a state or region, one must be sensitive to ethnic and other differences among applicator groups. In looking at the physical environment, we should also take a broad perspective. It may appear that applicators' attitudes must change when it would actually be easier to adjust the equipment. If a problem cannot be solved through garment redesign, perhaps it can be solved through package redesign.

Great strides have already been made in gaining better understanding of attitudes and practices regarding pesticides and protective clothing. However, farming still remains one of the more hazardous occupations, and pesticide contamination is still one of the hazards. Clearly, there is still much yet to be done.

ACKNOWLEDGMENTS

The author acknowledges the following for contributing support for work on this chapter: U.C. Agricultural Health and Safety Center at Davis, supported by NIOSH Cooperative Agreement Number U07/CCU 906162-01; North Central Regional Project NC-170: Reducing Pesticide Exposure of Applicators through Improved Clothing Design and Care.

REFERENCES

1. M. J. Clark, Research needs for assessing human health effects of pesticides; Proceedings of the 16th Annual ENR Conference, Illinois Department of Energy and Natural Resources, Springfield, Ill., 1987, pp. 81–90.

2. R. F. Wasserstrom and R. Wiles, *Field Duty: U.S. Farmworkers and Pesticide Safety*. World Resources Institute, Washington, D.C., 1985.
3. F. E. Allen, W. R. Grace going 'natural' with bug killer, *Wall Street Journal CCXVIII* (86):B1 (1991).
4. T. Namba, C. T. Nolte, J. Jackrel, and D. Grob, Poisoning due to organophosphate insecticides, *American Journal of Medicine 50*:475 (1971).
5. E. Kahn, Pesticide related illness in California farm workers, *Journal of Occupational Medicine 18* (10):693 (1976).
6. P. W. Smith, W. B. Stavinoha, and L. C. Ryan, Cholinesterase inhibition in relation to fitness to fly, *Aerospace Medicine 39*:754 (1968).
7. D. R. Metcalf and H. H. Holmes, EEG, psychological, and neurological alterations in humans with organophosphorous exposure, *Annals of the New York Academy of Science 160*:357 (1969).
8. E. P. Savage, T. J. Keefe, L. M. Mounce, R. K. Heaton, J. A. Lewis, and P. Burcar. Chronic neurological sequelae of acute organophosphate pesticide poisoning, *Archives of Environmental Health 43* (1):38 (1988).
9. D. P. Buesching and L. Wollstadt, Cancer mortality among farmers, *Journal of National Cancer Institute 73* (3):503 (1984).
10. L. F. Burmeister, G. D. Everett, S. F. Van Lier, and P. Isacson, Selected cancer mortality and farm practices in Iowa, *American Journal of Epidemiology 118* (1):72 (1983).
11. L. F. Burmeister, Cancer in Iowa farmers: Recent results, *American Journal of Industrial Medicine 18*:295 (1990).
12. S. K. Hoar, A. Blair, F. F. Holmes, C. D. Boysen, R. J. Robel, R. Hoover, and J. F. Fraumeni, Agricultural herbicide use and risk of lymphoma and soft-tissue sarcoma, *Journal of the American Medical Association 256* (9):1141 (1986).
13. D. D. Weisenberger, Environmental epidemiology of non-Hodgkin's lymphoma in eastern Nebraska, *American Journal of Industrial Medicine 18*:303 (1990).
14. K. Wiklund, J. Dich, and L.-E. Holm, Risk of malignant lymphoma in Swedish pesticide appliers, *British Journal of Cancer 56*:505 (1987).
15. B. M. Francis Health effects of pesticides, Proceedings of the 16th Annual ENR Conference, Illinois Department of Energy and Natural Resources, Springfield, Ill. 1987, pp. 65–79.
16. L. Mehler, S. Edmiston, D. Richmond, M. O'Malley, and R. I. Krieger, *Guide to the California Pesticide Illness Surveillance Program, 1988*, California Department of Food and Agriculture, Sacramento, Calif., 1990.
17. R. Mines and M. Kearney, *The Health of Tulare County Farmworkers*, unpublished manuscript, Tulare County Department of Health.
18. S. Vaupel, Health perceptions and health issues of Ventura County agricultural workers, paper presented in the U.C. Agricultural Health and Safety Center 1992 Seminar Series, Davis, Calif.
19. J. T. Griffiths, C. R. Stearns, and W. L. Thompson, Parathion hazards encountered spraying citrus in Florida, *Journal of Economic Entomology 44*:160 (1951).

20. G. E. Quinby and A. B. Lemmon, Parathion residues as a cause of poisoning in crop workers, *Journal of the American Medical Association 166* (7):740 (1958).
21. D. Culver, P. Caplan, and G. S. Batchelor, Studies of human exposure during aerosol application of malathion and chlorthion, *A.M.A. Archives of Industrial Health 13*:37 (1956).
22. H. R. Wolfe, Workers should be protected from pesticide exposure, *Weeds, Trees and Turf*, 12 (1973).
23. M. L. Rogers, Attitudes and practices of pesticide applicators regarding necessary and regular use of protective clothing and safety equipment, Proceedings of the National Conference on Protective Clothing and Safety Equipment for Pesticide Workers, Federal Working Group on Pest Management, Rockville, Md., Distributed by NTIS, U.S. Department of Commerce, 1972.
24. G. M. Beal, J. M. Bohlen, and H. G. Lingren, *Behavior Studies Related to Pesticides: Agricultural Chemicals and Iowa Farmers*, Special Report No. 49, Iowa State University, Ames, Iowa, 1966.
25. J. O. DeJonge, J. Vredevoogd, and M. S. Henry, Attitudes, practices, and preferences of pesticide users toward protective apparel, *Clothing and Textiles Research Journal* 2:9 (1983–84).
26. V. B. Keeble, M. J. T. Norton, and C. R. Drake, Clothing and personal equipment used by fruit growers and workers when handling pesticides, *Clothing and Textiles Research Journal 5*:1 (1987).
27. H. M. Perkins, E. M. Crown, K. B. Rigakis, and B. Eggertson, The attitude and behavior of farmers towards disposable protective clothing: An experimental field study, Proceedings of the First International Symposium on the Impact of Pesticides, Industrial and Consumer Chemicals on the Near Environment (B. M. Reagan, D. Johnson, and S. Dusaj, eds.), Kansas State University, Manhattan, Kan., (1988), pp. 234–242.
28. M. Rucker, D. Branson, C. Nelson, W. Olson, A. Slocum, and J. Stone, Farm families' attitudes and practices regarding pesticide application and protective clothing: A five-state comparison—Part 1: Applicator data, *Clothing and Textiles Research Journal 6*:37 (1988).
29. C. Nelson, M. Rucker, W. Olson, D. Branson, A. Slocum, and J. Stone, Farm families' attitudes and practices regarding pesticide application and protective clothing: A five-state comparison—Part 2: Launderer data, *Clothing and Textiles Research Journal* 2:36 (1988).
30. M. H. Rucker, K. M. McGee, and T. Chordas, California pesticide applicators' attitudes and practices regarding the use and care of protective clothing, *Performance of Protective Clothing*, ASTM STP 900 (R. L. Barker and G. C. Coletta, eds.), American Society for Testing and Materials, Philadelphia, 1986, pp. 103–113.
31. J. F. Stone, K. J. Koehler, C. J. Kim, and S. J. Kadolph, Laundering pesticide soiled clothing: A survey of Iowa farm families, *Journal of Environmental Health 48*:259 (1986).

32. J. F. Stone, M. L. Eichner, C. J. Kim, and K. J. Koehler, Relationship between clothing and pesticide poisoning symptoms among Iowa farmers, *Journal of Environmental Health 50* (4):210 (1988).
33. E. M. Rogers and F. F. Shoemaker, *Communication of Innovations*, The Free Press, New York, 1971.
34. *Protective Clothing for Pesticide Users*, National Agricultural Chemical Association, United States Department of Agriculture, and United States Environmental Protection Agency, Washington, D.C., 1986.
35. T. L. Lavy, J. D. Mattice, and R. R. Flynn, Field studies monitoring worker exposure to pesticides, *Pesticide Formulations and Application Systems* (K. G. Seymour, ed.), Amerian Society for Testing and Materials, Philadelphia, 1983, pp. 60–74.
36. G. W. Wicker, W. A. Williams, J. R. Bradley, and F. E. Guthrie, Exposure of field workers to organophosphorus insecticides: Cotton, *Archives of Environmental Contamination and Toxicology 8*:433 (1979).
37. J. M. Laughlin, C. B. Easley, R. E. Gold, and R. M. Hill, Fabric parameters and pesticide characteristics that impact on dermal exposure of applicators, *Performance of Protective Clothing* (R. L. Barker & G. C. Coletta, eds.), American Society for Testing and Materials, Philadelphia, 1986, pp. 136–150.
38. G. G. Choudray and G. R. B. Webster, Protocol guidelines for the investigations of photochemical fate of pesticides in water, air, and soils, *Pesticide Reviews 96*:79 (1985).
39. C. J. Kim, J. F. Stone, J. R. Coates, and S. J. Kadolph, Removal of alachlor residues from contaminated clothing fabrics, *Bulletin of Environmental Contamination and Toxicology 36*:234 (1986).
40. S. Helmers, J. Dykstra, and B. Kemp, Cholinesterase risk for Iowa farmers, *Iowa Medicine 80* (2):73 (1990).
41. M. Hussain, K. Yoshida, M. Atiemo, and D. Johnston, Occupational exposure of grain farmers to carbofuran, *Archives of Environmental Contamination and Toxicology 19* (2):197 (1990).
42. H. M. Perkins, *Attitudes and Behavior of Farmers Toward Disposable Protective Clothing: An Experimental Field Study*, unpublished master's thesis. University of Alberta, Edmonton, 1988.
43. K. B. Rigakis, S. Martin-Scott, E. M. Crown, N. Kerr, and B. Eggertson, Limiting pesticide exposure through textile cleaning procedures and selection of clothing, *Agricultural Forestry Bulletin 10* (2):24 (1987).
44. R. A. Fenske, A. M. Blacker, S. J. Hamburger, and G. S. Simon, Worker exposure and protective clothing performance during manual seed treatment with lindane, *Archives of Environmental Contamination and Toxicology 19*:190 (1990).
45. M. Rucker, C. Heffner, and K. York, Protective gloves for agricultural application of pesticides: A needs assessment, Proceedings of the First International Symposium on the Impact of Pesticides, Industrial and Consumer Chemicals on

the Near Environment (B. M. Reagan, D. Johnson, and S. Dusaj, eds.), Kansas State University, Manhattan, Kan., 1988, pp. 255–256.

46. C. Vlek and P. J. Stallen, Judging risks and benefits in the small and in the large, *Organizational Behavior and Human Performance 28*:235 (1981).

47. H. R. Wolfe, W. F. Durham, and J. F. Armstrong, Exposure of workers to pesticides, *Archives of Environmental Health 14*:622 (1967).

48. D. C. Staiff, J. E. Davis, and E. R. Stevens, Evaluation of various clothing materials for protection and worker acceptability during applicatio of pesticides, *Archives of Environmental Contamination and Toxicology 11*:391 (1982).

49. L. C. Shern, *Protective Clothing Actions of Michigan Corn and Apple Growers in Regard to Pesticides*, unpublished master's thesis, Michigan State University, 1986.

50. L. C. Shern and A. C. Slocum, Perceptions of risk and protective clothing actions of Michigan farm families, Proceedings of the First International Symposium on the Impact of Pesticides, Industrial and Consumer Chemicals on the Near Environment, (B. M. Reagan, D. Johnson, and S. Dusaj, eds.), Kansas State University, Manhattan, Kan., 1988, pp. 257–266.

51. C. D. Staubus, *Health Beliefs and Precautionary Clothing Practices of Oklahoma Farm Families in Regard to Pesticide*, unpublished master's thesis, Oklahoma State University, 1990.

52. A. Pryor, They know not what they do, *California Farmer 266* (6):6 (1987).

53. M. Fishbein and I. Ajzen, *Belief, Attitude, Intention and Behavior*, Addison-Wesley, Reading, Mass., 1975.

54. M. J. Ryan and E. H. Bonfield, Fishbein Extended Model and consumer behavior, *Journal of Consumer Research 2* (2):118 (1975).

55. I. M. Rosenstock, Why people use health services, *The Milbank Memorial Fund Quarterly 44* (3):94 (1966).

5

Refurbishing Pesticide-Contaminated Clothing

JOAN LAUGHLIN University of Nebraska—Lincoln, Lincoln, Nebraska

I. INTRODUCTION

Pesticides "protect us from insects, weeds, disease and hunger, but exposure to some pose a risk of cancer, birth defects, genetic mutations, and sterility" [1, p. 148]. Exposure does not mean the same as toxicity, but exposure is a problem for agricultural workers, urban pesticide applicators, emergency response teams, and those who work in pesticide manufacturing and distribution industries.

> Acute exposure to pesticides may result in systemic or local disease. With systemic poisonings, the clinical picture reflects the known toxicology of the compound and occurs shortly after exposure. Cholinergic illness due to cholinesterase inhibition from excessive organophosphate and carbamate exposure is the commonest type of systemic poisoning. Other less common, but equally life threatening examples of acute poisonings include the gastrointestinal, hepatitis, renal and pulmonary phases of paraquat poisonings, the metabolic stimulation that follows excessive exposure to the nitrophenol group of pesticides . . . and the seizure disorders that herald the excessive chlorinated hydrocarbon exposure" [2, p. 2].

Savage et al. [3] established a relationship between acute organophosphate poisonings occurring at a point in time and subsequent neurological problems. Indications of cause-and-effect relationships between pesticide use and human health, including cancer occurrences, are reported [4–7].

The importance of pesticide dermal exposure has not received sufficient attention until recently, perhaps because the older pesticides are absorbed through the skin to a lower extent than the pesticides used currently [8]. Batchelor and Walker [9], Durham and Wolfe [10], Gold et al. [11,12], and Leavitt et al. [13] used cotton gauze pads placed on the skin and clothing of workers to monitor worker exposure to pesticides during application. The collected residues are used to estimate total dermal exposure. Potential inhalation exposure also was monitored. For outdoor foliage applications, dermal exposure to pesticides is of greater concern than inhalation exposure. Dermal exposure accounts for 87–90% of the total human exposure to pesticides [14–16]. Wolfe [17] indicated that less than 1% of total exposure to pesticides is via respiration. The vast majority of occupational poisonings today are dermal in nature.

Agricultural workers and others may not be as aware of dermal exposure as they are of inhalation exposure [18]. Thus, reduction of pesticide dermal exposure among mixer/handlers and applicators has prompted widespread recommendations for protective clothing [12,19]. Dermal exposure could be reduced if workers wore moisture-resistant body coverings, but availability, cost, and comfort preclude widespread adoption of such apparel among pesticide applicators [20]. Agricultural workers, commercial landscape personnel, pest control operators, and aerial applicators are targeted occupations for intervention.

Current labeling of pesticides calls for the prudent use of protective clothing when mixing, loading, or applying these chemicals. Kansas farmers who did not use protective clothing and equipment when working with herbicides were found to have higher risk of cancer than those who did protect themselves [7]. Wicker et al. [21] reported that contaminated work clothing may act as an occlusive dressing that enhances dermal absorption of organophosphate insecticides. Lavy et al. [22] established that individuals wearing obviously contaminated clothing show greater levels of exposure than do other workers.

In general, the pesticide products used in agriculture, homes, and commerce are formulated products. Formulated products vary in concentration of the active ingredient (A.I.), as the technical material is diluted with inert carriers to a concentration appropriate for the intended use. Generally, a full-strength formulated product poses more significant problems than application dilutions, although for pesticides that are very high in toxicity, diluted mixtures may be just as great a concern.

Toxicity may vary by species, route of intake, formulation of pesticide, products' age or relative potency, and sex, age, and nutritional status of the recipient. Intoxication can occur orally, dermally, or through inhalation. The LD_{50} values (statistical estimates of the dosage fatal to 50% of a large population) apply to oral and dermal dosages. The LD_{50}s are expressed as milligrams of toxicant per kilogram of body weight. For pesticide labeling, signal words of *high toxicity*, *moderate toxicity*, and *low toxicity* indicate categories derived from the LD_{50}s. For example, three pesticides included in this discussion are methyl parathion, with an oral LD_{50} of 14 (for males) and a dermal LD_{50} of 67, is rated highly toxic; diazinon, with an oral LD_{50} of 108 and a dermal LD_{50} of 900, is rated moderately toxic; and malathion, with a dermal LD_{50} of 1,375 and an oral LD_{50} of greater than 4,444, is rated low in toxicity.

People respond to hazards as they perceive them. One of the difficulties with pesticides is that the hazard–benefit ratio favors benefits in the perceptions of users. The potential hazards of pesticides are deprecated by pesticides users, whereas the potential benefits of pesticides are of primary (economic) importance. Estimates are that as many as 300,000 farm workers in the United States may suffer pesticide-related illness each year. Savage et al. [23] estimated the rate of hospital-admitted pesticide poisonings to be 10 per 100,000 admissions.

Durable cotton overalls, a long-sleeved shirt, and trousers are minimum levels of protective clothing [24]. Agricultural workers report that pesticides get on their clothes during application; but most do not routinely use protective apparel. Fabrics commonly used for work clothing will continue to be worn by pesticide applicators. Accidental spills of concentrated or diluted pesticides may be absorbed by fabrics such as cotton and cotton/polyester blends. The Federal Task Group on Occupational Exposure to Pesticides [25] requires that "all protective clothing shall be thoroughly washed after each day's use . . . the clothing shall be adequately cleaned before it's passed on." After clothing has been laundered, it is generally considered to be clean by most applicators and their family members according to survey results of Stone et al. [26]. It is possible that when the wearer is not aware of pesticide residues in fabrics, dermal exposure might continue from wearing inadequately decontaminated garment. Thus, refurbishment is an essential and critical part of continued safety, and decontamination is essential if the garment is to be worn again.

A recent example is the multiple pesticide intoxication case attributed to residue in laundered apparel [27]. A worker in a pesticide-formulating plant became ill and required hospitalization. Within a 2-week period, two other workers required hospitalization for similar symptoms. Reconstruction of the incident shows that the first worker accidentally spilled methyl parathion on his coveralls. He took the required safety measures of an immediate shower, then placed the plant-issued uniform in a disposal bag to be burned. Instead, industry services laundered the uniform and reissued it to him, leading to a second episode of illness and hospitalization. While he was hospitalized, the launderers washed the coveralls; then reissued the garment to a second worker, who subsequently also required hospitalization for acute pesticide poisoning. After yet another laundering, a third worker wore the coveralls and then required medical attention for pesticide intoxication. The plant launched a thorough investigation of plant practices that may have contributed to the incident *before* discovering that the repeatedly, but inadequately, laundered coveralls rather than manufacturing procedures were responsible for the incidences of worker intoxication [27].

Southwick et al. [28] reported the death of an adult male attributed in part to wearing refurbished clothing with residues of methyl parathion. Young boys in Fresno, Calif., suffered parathion intoxication as a result of wearing contaminated-then-laundered blue jeans [29]. Bedsheets and blankets, laundered after transport with agricultural chemicals, were the source in another pesticide poisoning accident [30].

Historically, pesticide exposure has been monitored with cotton gauze fabric (alpha-cellulose pads) as collectors to determine ''safe'' reentry periods for crops treated with organophosphates [18,31]. Finley and associates [32–35] explored whether worker's clothing also served as pesticide collectors and whether residues could be removed during refurbishment. Finley and coworkers have established that residues remain after decontamination. The preponderance of decontamination is done at home using family or commercial laundering equipment. A complete removal of pesticide residues from laundered fabric seldom has been achieved [36]. In a case study with a farmer's coveralls worn through four planting seasons, Stone and Stahr [37] reported that residues of Treflan, Lorsban, Counter, Dyfonate, and Thimet encountered in the normal management of corn and soybean crops are not removed completely from clothing with home laundering and that the residues remain in fabric for an extended time (even as long as 4 years). Decontaminating protective apparel through refurbishment has been evaluated by multifactorial studies, yet a systematic accounting of interacting factors in a metasynthesis of findings is needed. This discussion will address that objective.

II. MECHANISMS OF SOILING AND SOIL REMOVAL

Research to date on removal of pesticide residues from work clothing has addressed a macroscopic examination of a complex system by using a reductionist approach of examining a set number of variables in a system. The work is based on an understanding of the mechanisms of soiling and soil removal, including the optimum balance of chemical energies, thermal energies, and mechanical energies in the refurbishment process [36]. Pesticide as soil in fabric brings to the laundering system an additional set of chemical and mechanical interactions that may parallel or may differ from usual soil in textiles. Pesticides as soil may consist of (1) solid particles, such as clay, chalk, or talc, as carriers of the active ingredient; (2) liquid emulsions, usually an oily media, as carriers; and (3) mixtures of solids and liquids. The extent of soil removal in the cleansing process is affected by the soil, fabric substrate, cleaning method, and interactions among all three factors. Kissa [38] has summarized the kinetics of soil removal processes as determined by the soil type: (a) the rate of water-soluble soil removal is proportional to soil concentration; (b) particulate soil removal is not constant because some soil particles are easier to remove than others; (c) oily soil release is complicated by several soil removal mechanisms occurring in stages: (1) induction period, during which the water and surfactant diffuse in to the soil–fiber interface and into the soil with a slow rate of soil removal; (2) separation of the soil from the fabric, where soil removal is usually not complete; and (3) a final period (leveling), during which soil removal is very slow, and some residual soil remains on the fabric.

The mechanism of soiling and difficulty in soil removal include penetration of the soil [dependent upon surface tension of soil and fiber, viscosity of the soil, and distance (capillary sorption) between fibers and interstices between yarns]; entrapment of particulate material in the structure of the fiber and/or in spaces of fibers fractured by the mechanical wear during use; and the chemical reaction of soil with fiber and finish [39]. The usual processes of soil removal from textile substrate are complicated by the chemical nature of pesticides as soil. Pesticides contain many chemical constituents, and each may present additional challenges in removal.

Work to date has examined the problem of ''invisible soil'' or pesticide residues remaining in refurbished protective work clothing [36]. In general, this body of work has examined processes of the soiling and soil removal mechanisms in refurbishment using multifactorial laboratory studies. From this work, laundering recommendations have been published [40] for minimizing direct and indirect exposure to workers and family members through laundering procedures that maximize residue removal.

III. PESTICIDES AS SOIL IN TEXTILES

Laughlin and Gold [36] have classified factors affecting successful decontamination as pesticide chemical class, pesticide formulation, and pesticide concentration. From additional research reports since that date, understanding of these factors has been increased, and solubility has been added to the list of factors affecting completeness of decontamination.

A. Pesticide Chemical Class

Lillie et al. [41,42] concluded that effective laundering procedures for reducing insecticide contamination may vary depending on insecticide type. This conclusion is based on work with diazinon (organophosphate), malathion (organophosphate), bromacil (uracil), propoxur (carbamate), and chlordane (organochlorine); however, the contamination level for each pesticide differed, and an insufficient number of pesticides was tested to observe the causal relationship of pesticide class on residue removal. In a study of 11 insecticides from the chemical classes of organophosphates (chlorpyrifos, diazinon, methyl parathion, dichlorvos, malathion, and dimethoate), organochlorines (aldrin, lindane, and chlordane) and carbamates (propoxur and carbofuran), Keaschall et al. [43] concluded that active ingredient may be a more reliable indicator of completeness of removal than chemical class (Figure 1). The basis for the conclusion includes the findings that differences in after-laundering residues were found both among chemical classes, with the largest overall residues for organochlorines and smallest for carbamates; and within classes, with greatest variability for organophosphates. As much as 37% of the organophosphate chlorpyrifos remains after one laundering. A summary paper [36] has drawn comparisons among pesticide chemical classes by compilation of the findings of numerous studies, and one study in progress employs a meta-analysis of work done at several state stations [44], but additional research needs to be done to understand the contribution of pesticide chemical class to residue removal.

B. Pesticide Formulation

In formulating the pesticide product, the technical material or active ingredient of the pesticide is formulated with a carrier. The carrier material is characterized through the terms used to describe formulations; for example, clay or talc as carrier for wettable powders and oils as carriers for emulsifiable concentrates. Removal of pesticide residues has been equated to removal of other soils in textiles, with parallels between particulate soils and wettable powders and oil soils and emulsifiable concentrates [45,46]. In a study of differences between the technical material (pesticide active ingredient) and formulated product for carbaryl and chlorothalonil, Fleeker et al. [47] reported that removal is greater for the technical material than the formulated

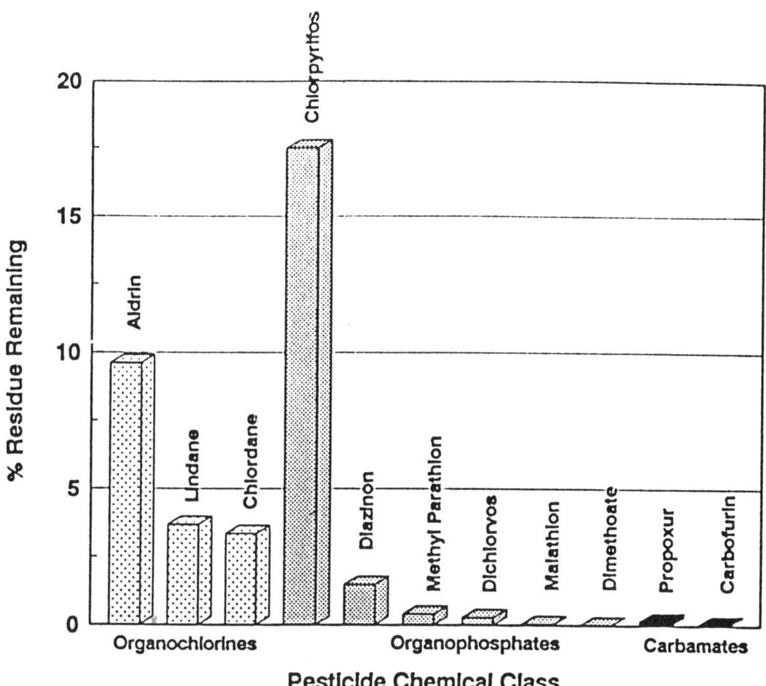

Figure 1 Differences among pesticide chemical classes in residue remaining after laundering. (Adapted from Ref. 43.)

product. Repeated decontamination in warm water is effective for the technical form of carbaryl, while the inert ingredients in the commercial formulation of chlorothalonil appear to increase the efficiency of aqueous extraction. Finley et al. [33] observed that methyl parathion is more completely removed when it alone soils the fabric; but, when methyl parathion is mixed with toxaphene and DDT, removal decreases by nearly 60%. Easter [46] considered Guthion to be an oily soil, and therefore difficult to remove from Gore-Tex fabric, and Captan to be a particulate soil, and therefore more difficult to remove from cotton fabric.

Laughlin et al. [48] concluded that removal is more complete for wettable powder and encapsulated formulations than for emulsifiable concentrate formulations of methyl parathion. In a study of 2,4-*D* herbicide, Easley et al. [49] supported the differences attributable to formulation, with water-soluble amine formulation more completely removed than the less soluble ester formulation. Fiber content of fabric is an interacting factor, concluded Easter and DeJonge [50] because azinphosmethyl, formulated as an oil-base concentrate, is more difficult to remove from synthetic fibers than from cotton. Captan, in a wettable powder formulation, poses more removal problems when cotton is the substrate than when synthetic fibers are the substrate [46]. There were no differences due to pesticide formulation (wettable powder versus flowable liquid) in a study of carbaryl and atrazine [51]. Raheel concluded that these two pesticides are readily removed from contaminated fabrics under commonly used decontamination regimens. But, Laughlin et al. [52] reported that the formulation effect is present between emulsifiable concentrate formulation and wettable powder formulation for each of two pyrethroids, cypermethrin and cyfluthrin.

Combinations of pesticides in one fabric are more difficult to remove in laundering than are the individual pesticides as single contaminants [33]. Toxaphene and DDT in combination with methyl parathion exhibited this combined effect, resulting in more methyl parathion residue than when methyl parathion is the only contaminate.

C. Pesticide Solubility

Braun et al. [53] concurred with Laughlin et al. [48] that decontamination efficiency is dependent upon formulation type but also proposed that solubility (in water) of the active ingredient is a factor in concert with formulation. They postulate that decontamination of Captan is highly effective due to its ease of degradation by hydrolysis rather than its solubility in water. Nelson and Fleeker [54] concluded that pesticide residue removal is not always a function of pesticide solubility. Water-insoluble chlorothalonil is more readily removed when formulated with "inert materials" than it is as unformulated product or technical material, while there is no difference between formulated and unformulated carbaryl in completeness of removal in decontamination through laundering. Carbaryl is moderately soluble but is readily hydrolyzed in alkaline media. Most detergent solutions are moderately to strongly alkaline (pH of 7.2 to 11.0). Formulation affects pesticide removal more than solubility.

Chiao-Cheng et al. [55] reported that all decontamination procedures, including the water-only extraction, were effective in removing more than 99% of carbamate (carbofuran and methomyl) insecticides from the cotton and the polyester fabrics. Given that carbofuran is unstable in alkaline media, the detergent solutes were alkaline, and surface deposit of the granular pesticides is more likely than deep penetration of liquid pesticide into the lumen, it was not surprising that carbofuran is more completely removed than methomyl. Generally, the water solubility of these carbamate insecticides would facilitate removal during decontamination; however, not all carbamates are similar in solubility.

The contribution of water solubility as well as alkaline hydrolysis on pesticide removal were corroborated by Raheel [51] in a study of carbaryl and atrazine. There are no significant differences in removal of the wettable powder or flowable liquid formulations of these pesticides.

Nelson et al. [56] provided evidence that paraquat removal is enhanced in a salt medium, such as that found in very hard water. They stipulate that this phenomenon may occur with other cationic pesticides such as difenzoquat and diquat. They conclude that predictions of pesticide decontamination based on solubility and formulation are problematic, since 2,4-*D* ester [49] and chlorpyrifos [57] are tenacious to decontamination, yet pesticides of similar structures are readily decontaminated.

D. Pesticide Concentration

Concentration of pesticide is a factor in decontamination through laundering. Easley et al. [58] reported a relationship between methyl parathion concentrations at spiking and after-laundering residue. Doubling the concentration of the pesticide (0.25% A.I., 0.50% A.I. 1.0% A.I. and 2% A.I.) decreases residue removal. The full (packaged) strength (54% A.I.) is more difficult to remove than the diluted (application) strength (1.25% A.I.). More than 80% of the concentrated product remains after one laundering, indicating that applicators need to use extreme caution when working with full-strength concentrates. Over two-thirds of the residue remains after ten complete laundering cycles, and the residue is approximately the level that was reported in laundered coveralls implicated in the repeated intoxication that contributed to the death of an adult male applicator in Utah [28]. A similar pattern is evident for paraquat [59] applied at 130 and 1,300 μg to fabrics before laundering.

Pesticide can "self-concentrate" in apparel fabrics if the garment is worn during repeated applications without refurbishment. Goodman et al. [60] reported that fabrics contaminated daily, then either laundered daily or only laundered after the 5 days of recontamination, show the accumulative build-up of methyl parathion (Table 1). The residue levels are much higher in the daily-spiked fabrics without daily laundering than in the duplicate fabrics laundered daily.

Table 1 Amount of Methyl Parathion on Fabric Specimen after Contamination and after Laundering for Specimens Spiked Daily

Day of treatment	Amount, μg/cm^2	Residue, μg/cm^2	Residue, %
1	11.68	0.25	2.14
2	30.84	1.20	3.89
3	45.37	5.44	12.00
4	53.08	6.43	12.14
5	53.42	7.09	13.28

Source: Adapted from Ref. 60.

IV. REFURBISHMENT FACTORS

Refurbishment is a detersive process. The effectiveness of the process is a result of the factors incorporated in the process: chemical energies of water, detergent, boosters, laundry aids, bleaches, and the like; mechanical energies of agitation; and thermal energies of temperature of wash cycles and rinse cycles, as well as time or length of cycles. Factors such as the soil saturation state of the textile and whether the soil is in liquid or particulate form make soil removal possible, or more problematic.

A. Chemical Energies

Soil removal in refurbishment is affected by the three interacting forms of energy—chemical, thermal, and mechanical. If one of these energies is diminished, the detersive action of the laundering process might be maintained at an optimum level through compensation from another energy. For example, an appropriate level of cleaning might be maintained even if the laundering temperature is lowered. To accomplish this optimum level of refurbishment at a lower temperature might require increased agitation (mechanical energy) or increased dosage or concentration of detergent.

Three components make up the detergent system (by which surfactants act on soil): solvent, solid, and surfactant. The solvent is usually water. The solids are distinguished by particle shape (e.g., powders or fibers), particle size, porosity, polarity, and chemical composition. Detergents are formulations of surface active agents, as essential constituents, and builders; boosters; fillers; and auxiliaries as secondary constituents. Surface active agents (surfactants) are chemical compounds that are preferentially absorbed by a solid–liquid or liquid–liquid interface. These surfactants are *amphiphilic*, that is, they have one or more hydrophilic groups and one or more hydrophobic groups [61]. Surfactants aggregate in solution at the interfaces between the surfactant–water solution and adjacent gases, liquids, or solids. This leads to formation of micelles in aqueous solutions and to structural surface films at the liquid–gas interface [62], with an increased concentration of surfactant at the air–liquid interface.

Four main types of surfactants are anionic, cationic, nonionic, and amphoteric. Of these, the anionic and nonionic types are included in the majority of detergents. The anionic surfactant's negatively charged component is sorbed onto the fabric surface by Van der Waals forces during laundering and is desorbed on rinsing. An anionic detergent will ionize in solution; however, the negatively charged ion is also attracted to the positively charged calcium and magnesium ions in hard water, which tend to reduce the surfactant's effectiveness. Nonionic surfactants exhibit marked surface activity at considerably lower concentrations than ionic ones with identical hydrophobic groups and have lower cmc (critical micelle concentrations) values [63]. In the detergency process, nonionic surfactants adsorb both on the fibers and on the particulate soil [62]. Correlations between adsorption and soil removal are dependent on the type of fabric as well as surfactant adsorption on the particulate soil.

The surfactants in detergent products vary in composition and are either single components or mixtures with technical products. Mixtures of different types of surfactants (e.g., anionic and nonionic) selectively adsorp. In a practical (use) concentration for cleaning or washing, the ratio of water molecules to surfactant molecules is approximately 20,000 + to 1. At this dilution, the surfactant has to accumulate or be adsorbed at sites to be effective.

Most published work of detersive action is with commercial products (mixtures) rather than homogeneous surfactants, because the isolated homogenous surfactant is unavailable and because the commercial detergent product is what is used in the home and commercial laundering process. Studies of pesticide residue removal in laundering also use this model of assessing the effectiveness of decontamination through laundering with commercial product.

1. *Detergent Type*

Statistical homogeneity is present for four detergents (nonionic heavy-duty liquid, anionics of phosphate, high phosphate, and carbonate) when methyl parathion contaminates the fabric [45]. Raheel and Gitz [64] observed little difference in effectiveness of a nonionic heavy-duty liquid detergent and a granular phosphate detergent in the removal of atrazine or carbaryl. Similarly, Chiao-Cheng et al. [55] reported no differences among three detergents—

granulated Tide and heavy-duty liquids All and Wisk—in removal of carbofuran and methomyl. Laughlin and Gold [65] found no detergent type (nonionic versus anionic) effects when laundering methyl parathion-spiked fabrics, and Parks et al. [66] confirmed this observation with methyl parathion on Gore-Tex fabric. Kim et al. [67] reported differences due to detergent with "light weight" fabrics, but not with "heavy weight" fabrics when alachlor is the pesticide applied to the fabrics. For the light weight fabrics, the residue remaining after laundering is twice as much when a heavy-duty liquid detergent (Wisk, surfactant not identified) is used at 0.2% v/v than when a commercial detergent (Tide, surfactant not identified) is used at 0.4% w/v or when a high-phosphate research detergent (AATCC 124, phosphate surfactant) is used at 0.2% w/v.

The notable exception to the consistent "no detergent type" effect comes when pesticide formulation is part of the experiments. Since the likelihood of removing particulate soils is increased when an anionic phosphate builder is included in the detergent formulation and since roll up of oily soils is enhanced with the higher surfactant level in nonionic heavy-duty liquid detergents, it is not surprising to see a formulation-dependent detergent effect. Emulsifiable concentrate formulations have an oil medium as a carrier, whereas wettable powder formulations have chalk, clay, or talc as a carrier. Laughlin et al. [52] reported a formulation–detergent type effect with cypermethrin and cyfluthrin in emulsifiable and wettable powder formulations (Figure 2) and recommend the detergent type should be selected with pesticide formulation in mind.

Of interest to pesticide applicators is the role of surfactant systems in the formulated pesticide products; surfactants are common constituents in application-strength pesticide dilutions. These surfactants, often cationics, are used to enhance attraction of the pesticide mixtures to the target foliar materials. Fleeker et al. [68] reported that when the unformulated chlorothalonil solution is diluted with 1% surfactant and placed on fabric specimens, the fungicide is more readily removed than the solution without the surfactant. The surfactant that is part of the pesticide dilution mixture actually improves the removal of chlorothalonil. Laughlin and Gold [69] see that residues of cationic surfactant (fabric softener) in the fabric before contamination or used in the laundering process after contamination (through 50 washing cycles) have no effect on methyl parathion absorption and retention.

2. *Detergent Concentration*

Hild et al. [70] found twice as much residue in fabrics decontaminated with half the usual amount of detergent and a leveling off in effectiveness of decontamination when twice the product manufacturer's recommended amount of detergent is used (Figure 3). Using the amount of detergent

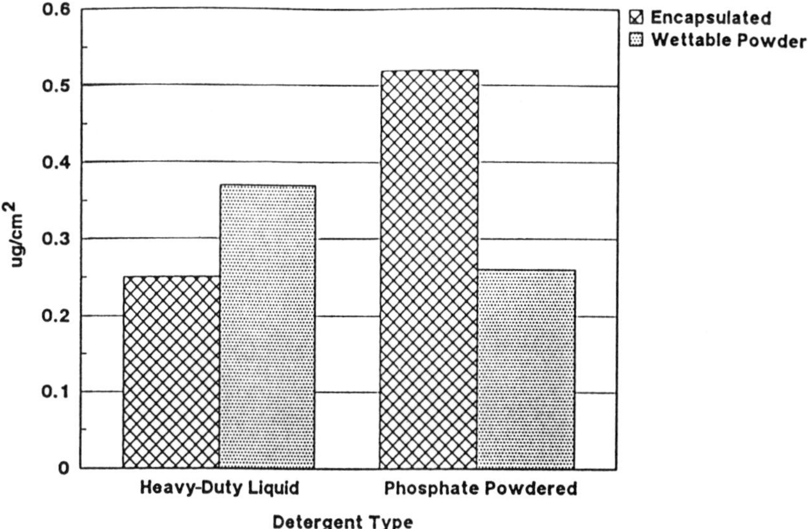

Figure 2 Contribution of detergent type and pesticide formulation to pyrethroid on after-laundering residues. (Adapted from Ref. 52.)

recommended on its label is not as effective as using one and one-half times the recommended dosage. They conclude decontamination should be done using 1.25 to 1.5 times the amount of detergent specified by the product's manufacturer.

3. *Prewash Product*

A prewash product may provide an additional dosage of surfactant or of surfactant and solvent. Without being diluted by mixing with the wash water, the concentration of surfactant on the soil enhances detersive action. Keaschall et al. [43] pointed out the improved efficacy of a prewash product plus detergent over decontamination with detergent only (for 11 pesticides studied). Deltamethrin, trifluralin, and trillate (field strength) are more completely removed when the laundry pretreatment product is used in the process than when it is not used [54,71]. Hild et al. [70], however, reported no difference in methyl parathion decontamination attributable to detergent type, perhaps due to the additional surfactant in the prewash product.

4. *Mineral Content of Water*

Much of the laboratory work has been done with distilled–deionized water, yet the presence of minerals in water may interfere with the cleaning effec-

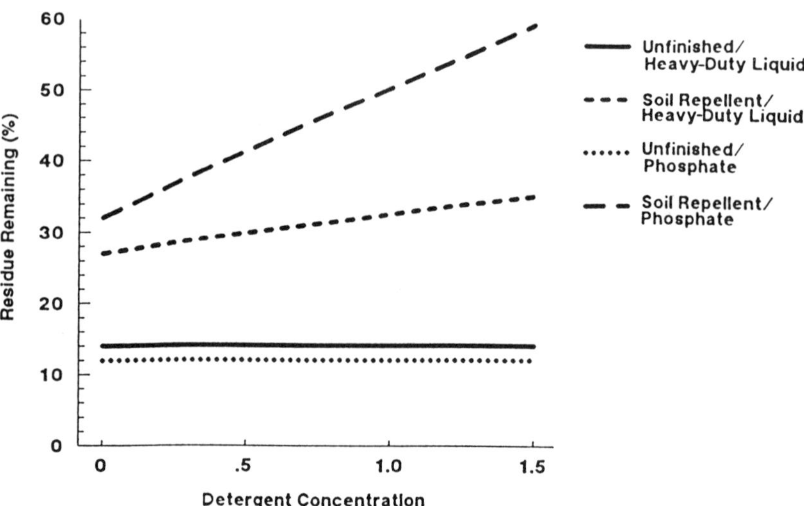

Figure 3 Contribution of detergent concentration and detergent type on after-laundering residue. (Adapted from Ref. 70.)

tiveness of detergent, the detersive action of an anionic product being reduced more than that of a nonionic product. Laughlin and Gold [72] reported that as level of water hardness increases above 600 ppm, the anionic phosphate detergent is less efficient in methyl parathion removal than the heavy-duty liquid detergent (Figure 4); however, use of a prewash product before laundering with the phosphate detergent results in residue levels similar to those produced with heavy-duty liquid detergents. With findings of a linear relationship, they concluded that as the level of water hardness increases by each 100 ppm, the methyl parathion residue retained after laundering the repellent-finished fabric increases approximately 0.55 $\mu g/cm^2$ and 2.67% (Table 2). However, Popelka [73] reported that water hardness has no effect on fonofos residues remaining after laundering. The work was done with "soft" water, and "tap" water, of approximately 150–200 ppm mineral content.

Laughlin [74] tested the contribution of thermal energy (temperature of washing) and chemical energy (detergent type, detergent concentration, use of a prewash product, and level of minerals in water) to methyl parathion residues remaining in fabrics after laundering. As mineral content of the water used in laundering increases, the percent of residue remaining after laundering increases (Figure 4). The one exception they noted is when a prewash product is included in the laundering protocol, thereby increasing the dosage of cleaning product. The prewash product's contribution to the laundering re-

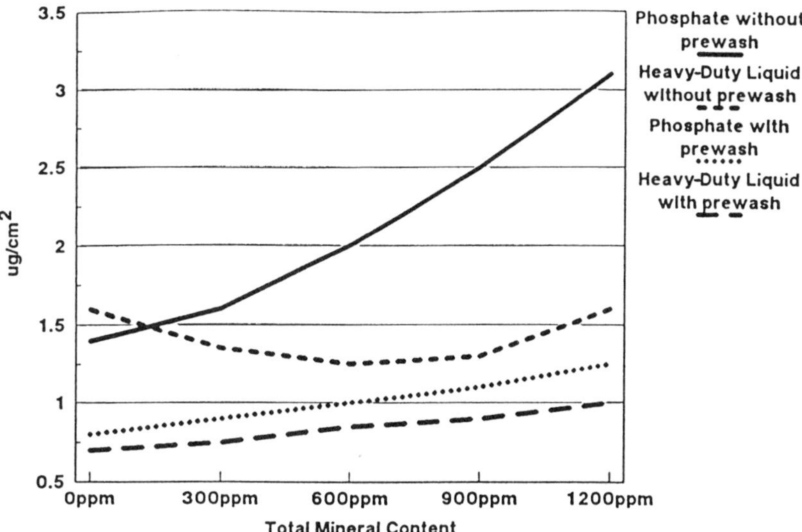

Figure 4 Contribution of mineral content of water, detergent type, and prewash product on after-laundering residue. (Adapted from Ref. 72.)

gime was more pronounced when a phosphate detergent is used than when a heavy-duty liquid detergent is used (Figure 5). Especially in the phosphate detergent wash, the additional aliquot of surfactant from the prewash product may counteract the impact of reduced chemical energy in a phosphate detergent regime as the phosphate surfactant interacts with the mineral content of water rather than with the soil in the fabric.

Table 2 Methyl Parathion Residues in Fabric Specimens Laundered with Detergent, with or without Prewash, and in Water that Varies in Mineral Content

Mineral content of water, ppm	Phosphate detergent amount, μg/cm^2	Phosphate detergent with prewash amount μg/cm^2	Heavy-duty liquid detergent amount, μg/cm^2	Heavy-duty liquid detergent with prewash amount, μg/cm^2
0	0.82	0.11	0.19	0.09
300	0.94	0.11	0.61	0.25
600	1.77	0.04	0.59	0.50
900	2.97	1.62	0.39	0.65
1200	2.32	1.35	0.15	0.25

(Adapted from Ref. 72.)

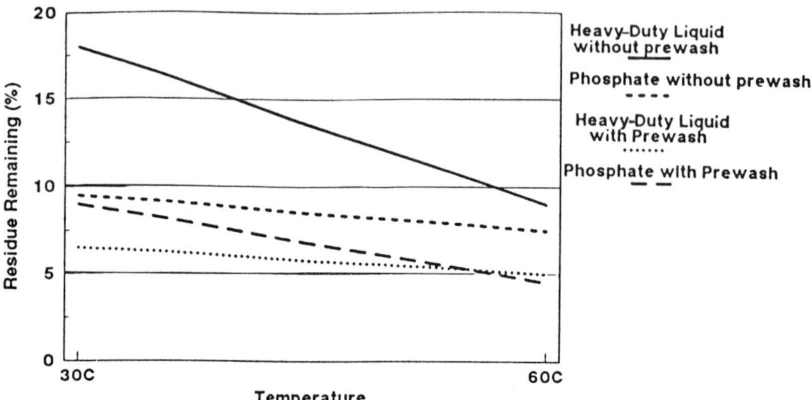

Figure 5 Contributions of prewash product to detergent's efficacy in lowering after-laundering residues at 30°C and 60°C. (Adapted from Ref. 74.)

Nelson et al. [56] observed the opposite effect when the pesticide of interest is paraquat. Salts, such as sodium chloride, in the hard water increase removal of paraquat. They report little difference in decontamination of chlorothalonil and carbaryl between distilled water and hard water. Adsorption studies of surfactants on inorganic salts and minerals have primarily been done with anionic and cationic surfactants. The lesser importance of nonionic surfactants may be attributed to the absence of electrostatic forces acting between the nonionic surfactant and the mineral surface [62].

5. *Laundry Auxiliaries*

Laundry auxiliaries can shift the chemical energy available in the refurbishment process. The shift can be to enhance or to diminish chemical energy, thereby contributing to, or detracting from, the efficiency of the laundering procedure. Laundering auxiliaries include fabric softeners, bleaches, and laundry boosters and starch, among others.

(a) Fabric Softeners. An important textile auxiliary often applied as a textile finish is a fabric softener that adds softness due to an interfiber lubrication effect. Fabric softeners are cation-active, surface-active compounds, usually amine salts, quaternary ammonium, or pyridinium derivatives. Quaternary ammonium salts are also used as emulsifying agents where absorption of the emulsifying agent onto the substrate is desirable (e.g., insecticidal emulsions). Given the lipophilic fabric softener, an emulsifiable concentrate pesticide may be miscible in this auxiliary.

Laughlin and Gold [69] reported that a single application of fabric softener does not affect chlorpyrifos absorption at initial contamination; however, a trend is present for increased after-laundering pesticide residue when fabric softener is introduced to the laundering process prior to contamination. Study of whether cationic surfactant (fabric softener) contributes to absorption and retention of methyl parathion residue after refurbishment led Laughlin and Gold [69] to conclude that at initial contamination, no consistent effect of previous decontamination with or without fabric softener exists for pesticide absorption. Greater after-decontamination residues occur on soil-repellent-finished specimens than on unfinished specimens. This held true for the comparisons among laundering cycles 1, 2, 3, 4, 5, and laundering cycles 10, 20, 30, 40, 50. Fabric softener—whether not used, used once, or used repeatedly—had no impact on residue remaining after decontamination.

(b) Starch. Using starch as a temporary textile finish to inhibit penetration through the fabric and then laundering the fabric to remove the starch, consequently removing the absorbed pesticide, was investigated by Sagan and Obendorf [75]. The higher the concentration of starch on textiles, the lesser the pesticide residue remaining after laundering; however, the use of amylase or an enzyme presoak to remove the starch did not increase decontamination. A 4% starch solution, applied as a temporary finish prior to exposure to pesticide, inhibits pesticide retention as compared to an unstarched fabric. This heavy loading of starch is not commonly used for work clothing. Laughlin et al. [57] reported no observable contribution of starch to completeness of chlorpyrifos decontamination when the starch was used at the more common 0.5% loading (Table 3).

(c) Bleach. Lillie et al. [42] and Laughlin et al. [48] observed only slight decreases in contamination of diazinon, chlordane, chlorpyrifos, bromacil,

Table 3 Contribution of Starch to After-Laundering Residues of Chlorpyrifos in Fabric Specimens.

Treatment before contamination	Cotton amount, μg/cm^2	Cotton/PET amount, μg/cm^2	Cotton residue, μg/cm^2	Cotton/PET residue, μg/cm^2
Laundered without starch	12.25	11.29	2.25	4.60
Laundered with starch			2.82	5.33

Source: Adapted from Ref. 57.

azinphosmethyl, and methyl parathion, when chlorine bleach is included in the washing process at temperatures between 30°C and 60°C. Current work at the University of Alberta indicates that liquid chlorine bleach (6.0% NaOCl), diluted to 0.04% with water at 50°C for a 3-hour soak, effectively reduces chlorpyrifos in the subsequent laundering procedure. This procedure does weaken the fabric, thus shortening its useful life. If this decontamination procedure is selected, launderers must be prepared for the strength loss over repeated decontamination cycles. It may be that the response is specific to chlorpyrifos; thus, the protocol needs to be applied to other pesticides before recommendations for bleach as a decontamination aid are made.

B. Thermal Energies

In the refurbishment process, thermal energy may be introduced in the presoak cycle, the wash cycle, or the rinse cycle. Generally, the impact of change in temperature is negligible in the presoak or rinse cycles and significant in the surfactant–wash cycle.

1. Washing Temperature

Kissa postulates that the wash temperature affects soil removal, through the formation of micelles from surfactant, dissociation of ionic surfactants into surfactant anions, adsorption of the surfactant on fiber and soil, and the viscosity of oily soil [38]. When washing cycle temperature is a variable, the importance of detergent surfactant type increases, due to the temperature dependence of surfactants. Higher temperatures of washing generally enhance soil removal. An increased wash temperature failed to establish a pattern of increased removal at increased temperatures in only one study. Kim et al. [67] reported that residues were more completely removed from the heavy weight fabrics at the "cold" temperature than at the "hot" temperature.

To determine whether thermal energy, temperature of washing, could compensate for diminished or enhanced chemical energy, Laughlin [74] examined detergent type, detergent concentration, and mineral content of water at water temperatures of 30°C and 60°C. She concluded that the hotter the water for washing, the lower the residues (Figure 5). The temperature effect was more noticeable for either detergent, without prewash product use. In other words, the elevated temperature is needed to compensate for the decreased chemical energy without the additional dosage of surfactant that would be present if a prewash product is included in the laundering process. At the lower temperature, there is similarity between the protocols with and without prewash product for the phosphate detergent washes.

C. Mechanical Energies

Fleeker et al. [47] and Nelson et al. [56] concluded that higher water temperatures, longer extractions time, and greater extraction volume increases the removal of carbaryl and chlorothalonil. Increasing the volume of water, thereby increasing the mechanical energy, improves the efficiency of paraquat removal in laundering [59].

Hild et al. [70] reported that significantly greater amounts of methyl parathion residue are found in fabrics decontaminated at one-quarter of the usual water volume (Table 4) (which, in effect, decreases the mechanical energy available in the laundering process), leading to recommendations to select maximum water volume (12 or 18 gal) for the laundering cycle. Water volume plays a more significant role in decontamination than does mechanical agitation, with no significant differences in the agitation levels in this particular study.

V. TEXTILE SUBSTRATE CHARACTERISTICS

The refurbishment of protective apparel and the detersive action in the laundering process is the product of three systems: a solvent (usually water), surfactant (usually detergent and laundry aids), and a solid (soil or textiles). The chemical composition of the textiles and the textile finishes, their fiber structure, factors such as size and porosity, yarn structure, and fabric structure have an impact on the level of cleaning from the refurbishment process.

A. Fiber Content

Polar fibers such as cotton and rayon interact very strongly with water, whereas nonpolar fibers such as polyester interact with water solely by dispersion forces. The success of refurbishment, in almost every combination of soil removal factors, is greater for cellulose than for hydrophobic nylon,

Table 4 Methyl Parathion Residue Remaining after Laundering Attributable to Water Level

Water level (x = full tub)	Amount, $\mu g/cm^2$	% residue remaining
$0.25x$	0.120	0.79
$0.50x$	0.071	0.53
$1.00x$	0.046	0.44
$1.50x$	0.050	0.47

Source: Adapted from Ref. 70.

polyester, and polytetrafluoroethylene [76,77]. Whether this holds true for pesticide soil is important to establish, to provide appropriate decontamination recommendations.

Obendorf and Solbrig [78,79] established, through scanning electron micrographs and x-ray diffraction analysis, that pesticide residue deposits in the lumen and crenulations of cotton and in the capillary spaces between polyester fibers. With pesticide residues deep within the fiber, aggressive treatment is needed to remove the residue.

Fiber content of fabric does not affect pesticide residue retention when the protective apparel fabrics are cotton or cotton/polyester blends in denims, chambrays, and uniform-weight fabrics [33,45,48,49,58,67, 80–86]. These observations hold true for methyl parathion, 2,4-*D* ester and amine, chlorpyrifos, fonofos, alachlor, and malathion. The exception is the work of Easter [46] and Lillie et al. [41,42]. Easter studied cotton and polyester denim and chambray, along with the barrier fabrics of uncoated Tyvek olefin, coated Tyvek olefin, and Gore-Tex nylon/polytetrafluoroethylene/nylon. She observed that wettable powder, Captan, is more difficult to remove from all-cotton fabrics than from the synthetics. Guthion emulsifiable concentrate is more difficult to remove from Gore-Tex than from other fabrics. Easter and DeJonge [50] concluded that fiber content–fabric type is the overriding factor in after-laundering residues.

Obendorf and colleagues [87–92] used electron beam x-ray microanalysis and backscattered electron images to study the distribution of soils, including pesticides on fabric surfaces, in yarn structures, on fiber surfaces, and within the fiber structure. Solbrig and Obendorf concluded that fiber content determines both the extent of penetration and the distribution of malathion across yarns and fibers. Malathion moves into the cotton fibers, to locations in the lumen and crenulations, as well as surrounding the fiber but does not penetrate into the polyester fiber. Obendorf and Solbrig [78,79] reported that methyl parathion reacts in similar ways, but with higher concentrations on the yarn surfaces of the cotton.

Lillie et al. [41] found that fiber content produces significantly different washability results, even though two fabrics of markedly different weights were used (cotton of 9.0 oz/yd^2, polyester of 2.8 oz/yd^2). These findings are generally contrary to the work of others. In fact, the data of Lillie et al. [41] show similar removal between the two fiber contents for carbaryl, prometron, and chlordane, with more diazinon residues removed from polyester than from cotton.

B. Yarn and Fabric Structure

1. *Yarn Structure*

Obendorf [90] reported that the location of multicomponent (clay and oil) soil is in the crevices between the closely spaced fibers and that laundering does remove large quantities of the oily and particulate soil from the yarn surfaces. Within the yarn bundle, the more deeply the soil has penetrated, the more difficult it is to remove in laundering. Work with malathion and methyl paration [79] established the parallels between response patterns for particulate/oily soils and for pesticides.

Kim and Kim [93] noted that the final location of pesticide (DDT) within textile substrate is related to fiber content and morphology, fabric geometry, and finish. Fiber irregularities serve as deposit sinks for the chemical soil. Understanding this phenomenon assists in understanding the difficulties in decontamination. Even though pesticide will fill the surface voids first, the more deeply it is allowed to penetrate into the fabric/yarn/fiber structure, the more difficult decontamination becomes.

2. *Fabric Weight*

Kim et al. [67] reported that fabric weight is an important factor in residue retention, with less retention in lighter weight fabrics and more residue retention in a heavier weight fabric; however, they evaluated only two fabrics, a light weight (blouse weight, 4 oz/yd^2) and a heavy weight (bottom weight, 14 oz/yd^2). They caution against light weight fabrics as it appears the fabrics might offer less protection. Olsen et al. [59] saw that the amount of paraquat remaining on the fabric after launderings appears to be a function of fabric weight. Laughlin et al. [52] compare two similar bottom weight fabrics, denims of 10.5 oz/yd^2 and 14 oz/yd^2, for pyrethroid residue remaining after decontamination. The lighter denim retains less pesticide than the heavier denim (Table 5). The heavier denim retains more of the emulsifiable concentrate formulation for both cyfluthrin and cypermethrin, whereas the lighter weight denim retains more wettable powder formulation for the cypermethrin, but not the cyfluthrin (Figure 6). These findings are consistent with the differences in soiling propensities of an oily soil and a particulate soil, the differences due to fabric weight in soil penetration, and the effectiveness of laundering in removing soil. The interrelationships of fabric weight with other pesticide–decontamination factors merits further investigation.

Table 5 Amount of Pyrethroid on 10.5 and 14.0 oz/yd^2 Fabric after Laundering

Treatment	Before laundering amount, μg/cm^2	After laundering amount, μg/cm^2	Percent residue remaining
10.5 oz/yd^2			
Cyfluthrin EC	2.1	0.37	17.6
Cyfluthrin WP	3.1	0.13	4.3
Cypermethrin EC	2.0	0.20	9.9
Cypermethrin WP	4.3	0.33	7.7
14.0 oz/yd^2			
Cyfluthrin EC	2.5	0.63	25.5
Cyfluthrin WP	3.1	0.32	10.4
Cypermethrin EC	2.3	0.55	23.9
Cypermethrin WP	2.8	0.26	9.4

Source: Adapted from Ref. 52.

C. Fabric Finishes

Soil-repellent finishes have been advocated to reduce absorption and penetration of pesticides through fabrics [94,95]. Fluorocarbon soil-repellent finishes on protective apparel fabric decrease pesticide absorption; however, these finishes may hinder pesticide removal in laundering [81]. Fluorocarbon polymers have very low surface tensions and, therefore, show very good oily soil repellency in air, but form high-energy surfaces in water during fabric laundering, disenabling hydrophobic soil dispersion in the aqueous wash medium [96]. Significantly greater percentages of pesticide residue remain in woven cotton–polyester fabrics that have a fluorocarbon finish compared to those that were unfinished [46,60,81]. Laughlin et al. [81] reported no significant differences in methyl parathion residues after laundering among unfinished, durable-press-finished fabrics, and soil-repellent-finished fabrics, even though the soil-repellent-finished fabrics had adsorbed only 20% of the loading with pesticide as the unfinished fabric. The low free surface energy of the fluorocarbon finish produces soil repellency and causes the fabric to be more repellent to the water–surfactant solutions. Goodman et al. [60] concluded that there are generally larger percentages of after-laundering residues on the soil-repellent-finished specimens than on the unfinished specimens; however, the initial amount of methyl parathion is less for the soil-repellent fabrics than for the unfinished fabrics.

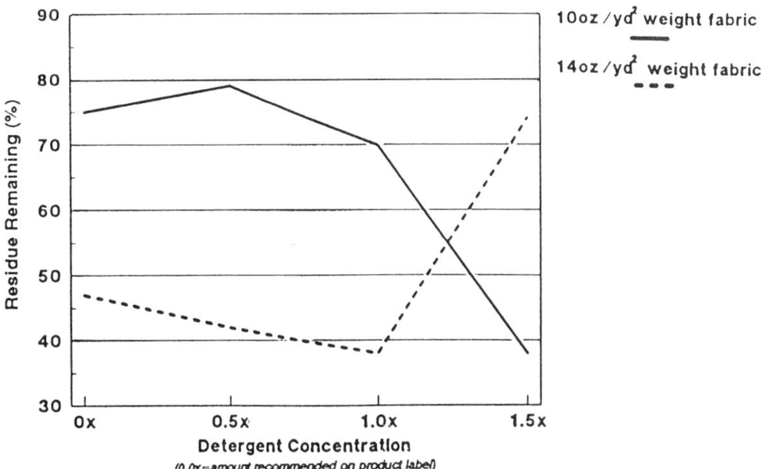

Figure 6 Contribution of fabric weight and detergent concentration to after-laundering residues of pyrethroids. (From Ref. 52.)

The most significant factor in pesticide contamination and removal by laundering is the soil-repellent finish (Figure 7) [52,57,69,72,81]. The soil-repellent finish inhibits absorption of chemicals on the specimens; however, any prior laundering diminishes the effectiveness of the consumer-applied soil-repellent finish [46,57,97]. Although residue remaining in fabrics after laundering is similar across unfinished fabrics, durable-press-finished fabrics, and soil-repellent-finished fabrics (through 50 laundering cycles) (Figure 8), it is important to note that the soil-repellent finish limits pesticide pickup at contamination to 80–90% of what is absorbed on unfinished fabrics [81]. After-laundering residue, as a proportion of initial contamination, is greatest for the soil-repellent-finished fabrics [46]. Obendorf and Solbrig [78] reported that the location of pesticide is selective per fiber content on a durable-press-finished (DMDHEU) cotton–polyester blend fabric. Higher concentrations of methyl parathion are on the surfaces of the polyester on the unlaundered unfinished fabric than on the durable-press fabric. Laundering reduces the amount of methyl parathion retained on a cotton–polyester blend fabric.

D. Barrier Fabrics

Barrier fabrics such as spunbonded and melt-blown fabrics are intended for limited use or disposable items, but some protective apparel information on

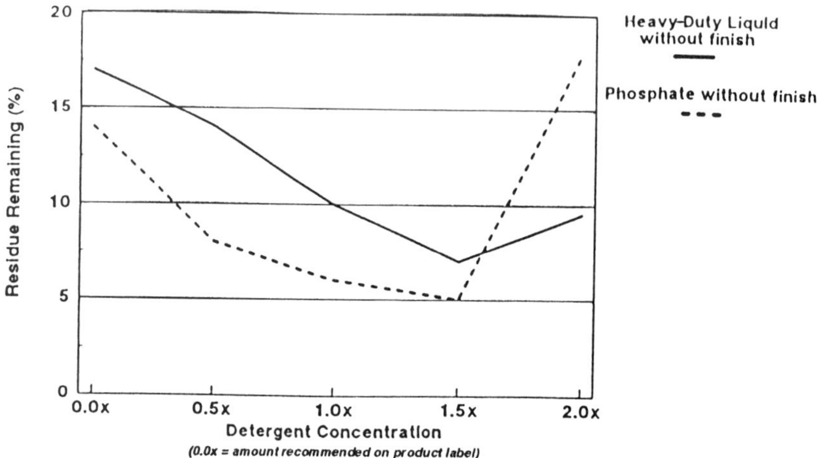

Figure 7 Contribution of fabric finish to after-laundering residues of methyl parathion. (Adapted from Ref. 74.)

packaging states that the fabrics can be laundered. Sontara fabrics offer the resistance to pesticide absorption, but this spunlaced nonwoven is not intended to be laundered and, in fact, does not survive one laundering cycle [98]. Fluorocarbon-repellent finishes limit absorption; however, laundering before spiking with the pesticide decreases the effectiveness of the functional finish in limiting pesticide pickup. Corona-treated spunbonded nonwoven fabrics maintain less adsorption through three laundering cycles. The functional finished fabric exhibits absorbency similar to the unfinished fabrics. Clearly laundering reduces the repellency of the nonwoven fabrics and should be avoided; however, there is no contribution of fluorocarbon finish to increased amounts of residue remaining in the nonwoven fabrics as is observed for woven cotton and cotton blend fabrics (Table 6) [98].

Alternatives to usual woven textiles are recommended for personal protective equipment for pesticide applicators. Since elevated basal temperatures and greater perceived thermal discomfort are reported for individuals wearing coated spunbonded ensembles, study of alternative–experimental fabrications is warranted [99]. Branson and Rajadhyaksha [100] studied decontamination of malathion from Gore-Tex, a multilayer fabric including a polytetrafluoroethylene (PTFE) membrane. A fairly high percent of pesticide is retained by the Gore-Tex; this is surprisingly so, since malathion is highly water soluble. Following decontamination, small concentrations of malathion are present on the surface of nylon fibers for both the face and back fabric, but

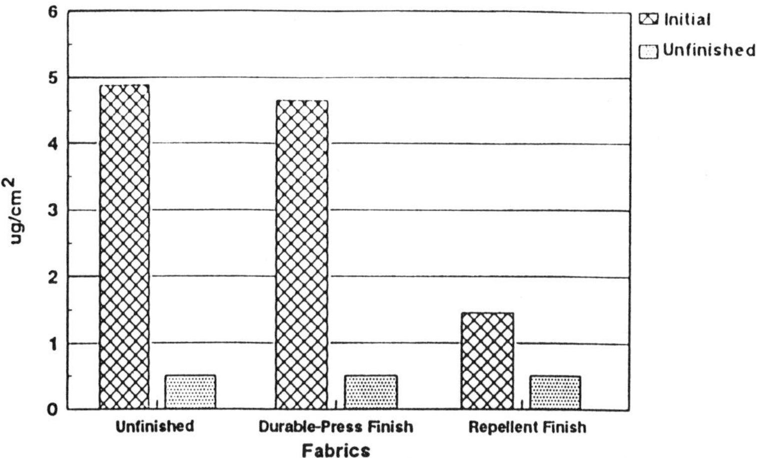

Figure 8 Contribution of fabric finish to initial and after-laundering residues of methyl parathion. (Adapted from Ref. 81.)

Table 6 Methyl Parathion Residues on Laundered Nonwoven Fabric as a Proportion of Contamination Before Laundering

Laundering cycle	Tyvek 1442 A (unfinished) residue remaining, %	Tyvek 1442R (repellent finished) residue remaining, %
0	8.42	3.61
1	5.67	2.39
2	5.53	2.70
3	3.96	3.16
4	4.54	4.67
5	8.43	4.52

Source: Adapted from Ref. 98.

malathion is trapped in the PTFE membrane, and the interstices between the membrane is the most difficult layer to decontaminate.

E. Fabric History

1. *Abrasion*

The mechanisms whereby a fabric becomes soiled and the difficulty in removal of soil from a textile fabric are enhanced by the changes in fiber, yarn, and fabric structure that occur during wear or during refurbishment. Soiling is enhanced by maximum surface area per unit volume; that is, the more crevices and fiber disruptions, the greater the potential for penetration and entrapment of soil. Abrasion is a usual part of wear and laundering of textiles as they are used. Laughlin and Gold [83] report that fabrics are altered by laundering and abrasion, such that they differ in methyl parathion soil absorption and completeness of residue removal in decontamination. Durable-press cotton is particularly susceptible to abrasion and to pesticide residue retention, due to the fiber fractures and fissures that develop during use and laundering and, thus, should not be selected for protective apparel. Acrylic acid soil release finish does not contribute to residue removal but does enhance soiling with pesticides and does inhibit pesticide removal.

2. *Previous Soiling*

With few exceptions, laboratory laundering has been used to duplicate the home laundering procedures, and the fabrics studied have been unused or "new" fabrics. Fabric that is used by pesticide handlers, mixers, and applicators may be soiled by body oils and by oily soils associated with equipment operation. Oily soils (including many pesticides) have significantly lower surface tensions and can penetrate fibers more readily than water-based soils. Obendorf et al. [87–92] reported that the number of washings plays a role in the amount of sebum retained by the fabrics. As oily soiled fabrics age, removal through laundering becomes increasingly difficult [101]. Particulate soil does not appear to adhere to the cotton or polyester but rather to an organic film that coats the fibers [77].

Obendorf and Klemash [91] found large amounts of residual triolein in the interfiber capillaries within the yarns and in the crenulations and the lumen of cotton. No triolein is located in the interior region of polyester fibers, but there are large quantities of residual oil on the fiber surfaces. A prewash product with organic solvent removes soil from surfaces of cotton. Weglinski and Obendorf [89] observed that the melting point of the oily soil significantly affects removal. Oleic acid and triolein (melting point $< 38°C$) are easier to remove than stearic acid and tristerin (melting point $> 38°C$).

Work on laundering pesticide-contaminated apparel fabric has established the relationship between emulsions and oily soils and between wettable powders and particulate soil [45,46]. Study of the relationships between oil soils (such as sebum or vegetable oils) and pesticide absorption and retention following refurbishment is important, since other studies only deal with "new" or unused fabrics [70]. In a study to determine whether the oily soils (synthetic sebum and vegetable oil) applied onto fabric specimens and then laundered prior to contamination contribute to pesticide absorption and retention, Laughlin and Gold [102] pursued interaction of these factors with a prewash laundry product.

Cotton–polyester specimens with a soil-repellent finish absorb less methyl parathion at initial contamination. Laundering removes significant quantities of pesticide; however, when the "history" of the specimen includes oily soil followed by laundering, unfinished specimens absorb less pesticide at contamination than do nonoily soiled specimens, whereas repellent-finished specimens absorb more methyl parathion. Oily soil, particularly on soil-repellent fabrics, can predispose to pesticide contamination.

Laundering removes significant amounts of chemical. Interactions of fiber content, finish, and use of prewash product and laundering treatment reveals that residues are always lower when the prewash product is used, but this product is not as effective on the all-cotton fabric as it is with the blend (Table 7). When oily soil is applied over unlaundered contaminate, more residue is removed in subsequent laundering. But, if oily soil is present before contamination, methyl parathion removal during laundering is less complete. These observations are consistent between two oily soils—synthetic sebum and vegetable oil.

Laughlin [103] evaluated the contribution of oily and particulate soil residue to pesticide residue removal using artificially soiled 100% cotton and 65% polyester–35% cotton fabrics. Initial methyl parathion contamination is not dependent on the soil level or fiber content of the fabric. Residues remaining after laundering are affected by soiling level. Pesticide residues are greater when the fabric has a heavy soil buildup, even though the initial contamination is lower. Based on these findings, protective apparel should be kept as clean as possible, with daily laundering, for the presence of soil residue affects decontamination of the fabrics.

Table 7 Methyl Parathion Residues Retained after Laundering by Fabric Specimens Soiled with Vegetable Oil

Treatment	After laundering amount, $\mu g/cm^2$	Percent residue remaining
Cotton without prewash	2.11	12.93
Cotton with prewash	0.79	7.21
Cotton/polyester without prewash	0.68	6.63
Cotton/polyester with prewash	0.32	4.21

Source: Adapted from Ref. 102.

Leveling refers to the very slow and incomplete removal of soil in laundering. Indicative of this leveling phenomenon is the retention of pesticide residues after refurbishment regardless of decontamination temperatures, detergent type, laundry additives, prerinse or wash cycle, prewash treatment, fiber content of fabric, textile finish, yarn size, or weave of fabric [36,37,43–49,51–60,64–72,75,78–92,97]. Also occurring during the decontamination process is the desorption of soil from the fabric via the washing liquor and redeposition on the same fabric or deposition onto another fabric [81]. Pesticide soiling occurs not only when the textile is worn during application but also during the laundering process [65,81]. Dispersion of pesticide from the original area of contamination to the entire fabric through soil redeposition has been reported [59,65].

VI. ALTERNATIVE DECONTAMINATION

The preponderance of the work to date on decontaminating pesticide applicator clothing assumes wet cleaning. Not all garments worn in exposure situations can be wet cleaned (e.g., turnout coats of volunteer fireman who might respond to an plane crash involving aerial application). Some pesticides respond to alternative solvents; others are affected by heat and ultraviolet light.

A. Dry Cleaning

Fleeker et al. [47] studied dry cleaning as an alternative to wet cleaning in decontamination. They reported that a hydrocarbon-based dry cleaning solvent was ineffective in removing carbaryl and chlorothalonil, whereas a perchloroethylene-based solvent removes greater than three-fourths of these pesticides. However, they caution that since dry cleaning solvents can transfer pesticide from contaminated garments to those not contaminated, even after removal of the originally contaminated apparel item, dry cleaning is not recommended. Ringenberg et al. [97] reach the same conclusion for chlorpyrifos-contaminated fabric (Table 8).

B. Volatilization

Laughlin and Gold [104] reported significant reduction in methyl parathion contamination in textiles held for periods up to 6 months, but with no decontamination. They suggest that storage before or after wet cleaning assists in minimizing the contaminant, through decomposition or volatilization of the chemical. Since only one-third of the concentrated (54% AI) methyl parathion is removed from fabrics through ten extractions, storage for

Table 8 Amount of Chlorpyrifos Transferred to Uncontaminated Fabric Specimens Through Concomitant Dry Cleaning with Contaminated Fabric Specimens

Treatment	Cotton	50% Cotton–50% PET
Initial amount, μg/cm^2	15.87	16.56
Amount transferred, μg/cm^2	0.005	0.002
Percent transferred	0.032	0.012

Source: Adapted from Ref. 97.

periods of time and/or at temperatures that maximize the loss of the chemical may be an alternative.

Generally, larger percentages of residue remaining after decontamination are found for the soil-repellent-finished specimens; however, the amount of initial contamination is lower for finished than for unfinished specimens [105]. Specimens that are contaminated, wet cleaned, and then held for up to 4,032 hr show lowest residues. Both conditions of holding and time affect residues, with greatest residues in specimens held at 0°C, and the smallest residues after the longest time periods. Although some decrease in residue is noted after 168 hr, significant declines are at 720 and 4,032 hr. The specimens held in ambient air require the extended time (4,032 hr) before a significant decline in residue retention is noted. Conditions of moving air and high humidity enhance volatilization or dissipation of methyl parathion, regardless of fiber content or finish. Wet cleaning plus holding in moving air significantly reduces the concentrated (54% AI) methyl parathion in fabric specimens.

The holding condition that contributes most to dissipation of methyl parathion residue is the presence of moving air at room temperature. Based on these data, if decontamination is not feasible for certain items of protective apparel, holding for periods in excess of 4 weeks and preferably over 24 weeks at room temperature and in moving air can be pursued. Storage of laundered garments at 20°C with moving air at higher humidity levels will assist in dissipating residues. Fabric finish plays a twofold role: the functional finish limits initial contamination and it is an inhibitor to residue removal during laundering.

C. Light and Heat

Branson and Rajadhyaksha [100] hypothesized that exposure to simulated sunlight, heat, and humidity might be an effective means of decontaminating fabrics since pesticides break down in the natural environment. When Gore-

Tex, a multicomponent fabric, including a PTFE film, is contaminated with malathion, no difference attributable to ultraviolet light, heat, or humidity is noted. Findings are consistent whether wet-cleaned or non-wet-cleaned fabrics are treated with the environmental conditions.

Kim et al. [106,107] studied the effects of heat in fabric drying on degradation of alachlor. Alachlor degrades to a trace level when exposed to heat of 150°C for 60 minutes or 200°C for 15 minutes. The microwave intensities of high, medium, and low settings are inadequate to produce parallel degradation.

VII. CONCLUSIONS

The perspective of a meta-analysis brings the mechanisms of soiling and soil removal from textiles to the problems of decontaminating protective clothing. The current work attempts to provide explanatory conclusions for residue retention, with the objective that these can lead to recommendations for maximizing decontamination. A caution about generalizations. Some pesticides produce very idiosyncratic responses to decontamination. An example is the paraquat/salt response. Other chemical control agents exhibit noticeable (and unique) responses to a highly alkaline medium or to bleach or are quickly volatilized. Responses such as these do not apply to *other* pesticides' response to decontamination. Given this caution, there are soil, substrate, and solvent responses that do maximize residue removal.

1. Geography of the soiling site is important. If the pesticide penetrates deeply into the structure of the fiber (e.g., cotton) or if all available soil "sinks" are filled with soil, decontamination is more difficult. This may occur when the pesticide is in a liquid form, when the pesticide is highly concentrated, when the pesticide repeatedly soils the fabric, or conversely, when a previous soil (particulate soil or oily soil) prevents the pesticide from penetrating deeply into the textile substrate. In the first instance, decontamination will be more problematic; in the second instance, decontamination may be facilitated because the pesticide is more available to be subjected to decontamination procedures.
2. The interactions of thermal, chemical, and mechanical energies are important to recommendations. If the chemical energy is reduced by mineral content of water, then increasing the thermal energy by raising the temperature of washing can compensate. Conversely, if thermal energy is reduced, then increasing the chemical energy through an additional dosage of surfactant, either adding a prewash product or increasing the concentration of detergent, can compensate.

3. Competing forms of chemical energy complicate recommendations. Increasing the detergent concentration may be counterproductive, if the detergent concentration is more than two times the concentration recommended by the detergent manufacturer. The effectiveness of anionic surfactants, used in water with elevated mineral content, is contingent on supplemental chemical energy, prewash product, or increase in the detergent concentration.
4. Formulation of pesticide is particularly important. Some formulations are more highly water soluble than other formulations. Wettable powder formulations respond as particulate soil and are more completely removed when a powdered phosphate or carbonate detergent is used, whereas emulsifiable formulations, with an oily medium as carrier, respond more to heavy-duty liquid detergents.
5. In general, decontamination of pesticides in a laundering process that includes a prerinsing cycle is more effective than one that does not include prerinsing. Use of prewash product, with a double exposure to surfactant system, enhances decontamination.
6. Fluorocarbon-repellent finishes will decrease initial pesticide absorption significantly, but they render the textile substrate more resistant to the detersive action of the surfactant system. Nonetheless, the initial repellency is an overriding factor in providing barrier properties to textiles.
7. Generally, vigorous decontamination procedures are more effective. Higher water temperatures, longer extraction times, increased agitation, and larger extraction volumes reduce residues remaining after decontamination.
8. Alternative decontamination methods may have value but merit further investigation before generalizations should be made. Some work has been done using volatilization (instead of wet cleaning), elevated heat, ultraviolet light, or dry cleaning. Responses may be pesticide specific, and the practical aspects of some methods may limit applicability (e.g., volatilization by hanging the garment in moving air for 6 months as an alternative to wet cleaning); yet, there may be instances where this is the only practical decontamination procedure.

Providing adequate protection to all who work with pesticides will continue. The greatest immanent health risk from pesticides comes through primary exposure. By making persons exposed to pesticides more aware of dangers, how to avoid contamination, and how to deal with contaminated clothing, the potential for adverse effects may be reduced. Given the 1,500 active ingredients, 35,000 formulations, and the hundreds of activities, sites, and exposure patterns, more work of a predictive nature [108] on successful refurbishment decontamination is needed.

REFERENCES

1. A. A. Boraiko, The pesticide dilemma, *Natl. Geographic 157*:145 (1980).
2. J. E. Davies, Health effects of global pesticide use. *World Resources Institute Report 2*, 1985, pp. 1–93.
3. E. P. Savage, T. J. Keefe, L. M. Mounce, R. K. Heaton, J. A. Lewis, and P. J. Burcar, Chronic neurological sequelae of acute organophosphate pesticide poisoning, *Archives of Environ. Health 43*:38 (1988).
4. D. P. Buesching and L. Wollstadt, Cancer mortality among farmers, *J. Natl. Cancer Institute*, *72*:503 (1984).
5. L. F. Burmeister, G. D. Everett, and S. F. Van Lier, Selected cancer mortality and farm practices in Iowa, *Amer. J. Epidemiology 118*:72 (1983).
6. K. P. Cantor, Farming and mortality from non-Hodgkin's lymphoma: A case-control study, *Internatl. J. Cancer*, *29*:239 (1982).
7. S. K. Hoar, A. Blair, F. F. Homes, C. D. Boysen, R. J. Robel, R. Hoover, and J. F. Fraumeni, Agricultural herbicide use and risk of lymphoma and soft-tissue sarcoma, *J. Amer. Medical Assoc.*, *256*:1141 (1986).
8. F. Matsumura and B. V. Madhukar, Exposure to insecticides, *Pharmacology and Therapeutics 9*:27 (1980).
9. G. S. Batchelor and K. C. Walker, Health hazards involved in use of parathion in fruit orchards of North Central Washington, *AMA Arch. of Industrial Hygiene 10*:522 (1954).
10. W. F. Durham and H. R. Wolfe, Measurement of urban applicators to carbaryl, *Bull. World Health Organ. 75*:75 (1962).
11. R. E. Gold, J. R. C. Leavitt, T. Holcslow, and D. Tupy, Exposure of urban applicators to carbaryl, *Arch. Environ. Contamin. Toxic. 11*:63 (1982).
12. R. E. Gold, T. Holcslow, D. Tupy, and J. B. Ballard, Dermal and respiratory exposure to applicators and occupants of residences treated with dichlorvos, *J. Economic Entomology 77*:430 (1984).
13. J. R. C. Leavitt, R. E. Gold, T. Holcslow, and D. Tupy, Exposure of professional pesticide applicators to carbaryl, *Arch. Environ. Contamin. Toxic. 11*:57 (1982).
14. H. R. Wolfe, J. F. Armstrong, D. C. Staiff, and S. W. Comer, Exposure of spraymen to pesticides, *Arch. Environ. Health 25*:29 (1972).
15. R. E. Gold and T. Holcslow, Dermal and respiratory exposure of applicators and residents to dichlorvos-treated residences, *Dermal Exposure Related to Use*, American Chemical Society Symposium Series #273, pp. 253, 1985. Washington, D.C.: ACS.
16. H. L. Maibach, R. J. Feldman, T. W. Milby, and W. F. Serat, Regional variation in percutaneous penetration in man, *Arch. Environ. Health 23*:208 (1971).
17. H. R. Wolfe, Field exposure to airborne pesticides, *Air Pollution from Pesticides and Agricultural Processes* (R. E. Lee, Jr., ed.), CRC Press, Cleveland, 1976, p. 157.

18. H. R. Wolfe, J. F. Armstrong, and W. F. Durham, Pesticide exposure from concentrated spraying, *Arch. Environ. Health*, *13*:340 (1966).
19. D. L. Gunn and J. G. R. Stevens, eds., *Pesticides and Human Welfare*, Oxford University Press, Oxford, England, 1976.
20. M. S. Henry. *Users' Perceptions of the Attributes of Functional Apparel*, unpublished master's thesis, Michigan State University, 1980.
21. G. W. Wicker, W. A. Williams, and F. E. Guthrie, Exposure of field workers to organophosphorus insecticides: Sweet corn and peaches, *Arch. Environ. Contamin. Toxic. 8*:175 (1979).
22. T. L. Lavy, J. D. Mattice, and R. R. Flynn, Field studies monitoring worker exposure to pesticides, *Pesticides Formulation and Application*, (K. G. Seymour, ed.) American Society for Testing and Materials, Philadelphia, STP #745, 1983, p. 60.
23. E. P. Savage, T. J. Keefe, H. W. Wheeler, and L. J. Helwic, *Natl. Study of Hospitalized Pesticide Poisonings, 1974–1976*, Epidemiologic Pesticide Studies Center, Colorado State University, Fort Collins, 1980.
24. Occupational Safety and Health Admin., *Fed. Register 38*:10715 (May 1, 1973).
25. Federal Register, *Emergency temporary standard for exposure to organophosphorus pesticides*, U. S. Government Printing Office, Washington, D.C. (May 1, 1973).
26. J. F. Stone, K. J. Koehler, C. J. Kim, and S. J. Kadolph, Laundering pesticide soiled clothing: A survey of Iowa farm families, *J. Environ. Health 48*:259. (1986).
27. N. J. Clifford and A. S. Nies, Organophosphate poisoning from wearing a laundered uniform previously contaminated with parathion, *J. Amer. Med. Assoc. 262*:3035 (1989).
28. J. W. Southwick, H. D. Mecham, P. M. Cannon, and M. J. Gortatowski, Pesticide residues in laundered clothing, *Proceedings of 3rd Conference of Environmental Chemicals and Human and Animal Health*, Fort Collins, Colorado State University, 1974.
29. M. C. Warren, J. P. Conrad, J. J. Bocian, Jr., and M. Hayes, Clothing-bourne epidemic. Organic phosphate poisoning in children, *J. Amer. Med. Assoc. 184*:94 (1963).
30. L. S. Anderson, D. L. Warner, J. E. Parker, N. Blumen, and B. D. Page, Parathion poisoning from flannelette sheets, *Can. Med. Assoc. J. 92*:809 (1965).
31. G. R. Hayes, A. J. Funckes, and W. V. Hartwell, Dermal exposure of human volunteers to parathion, *Arch. Environ. Health 8*:829 (1964).
32. E. L. Finley and J. R. Rogillio, DDT and methyl parathion residue found in cotton and cotton-polyester fabrics worn in cotton fields, *Bull. Environ. Contamin. Toxicol. 4*:343 (1969).
33. E. L. Finley, G. I. Metcalfe, F. G. McDermott, J. B. Graves, P. E. Schilling, and F. L. Bonner, Efficacy of home laundering in removal of DDT, methyl

parathion and toxaphene residues from contaminated fabrics, *Bull. Environ. Contamin. Toxicol. 12*:268 (1974).
34. E. L. Finley, J. M Bellon, J. B. Graves, and K. L. Koonce, Pesticide contamination of clothing in cotton fields, *LA Agric. 20*:8 (1977).
35. E. L. Finley, J. B. Graves, F. W. Hewitt, H. F. Morris, C. W. Harmon, F. A. Iddings, P. E. Schilling, and K. L. Koonce, Reduction of methyl parathion residues on clothing by delayed field re-entry and laundering, *Bull. Environ. Contamin. Toxicol. 22*:598 (1979).
36. J. M. Laughlin and R. E. Gold, Cleaning protective apparel to reduce pesticide exposure, *Rev. Environ. Contamin. Toxicol. 101*:94 (1988).
37. J. F. Stone and H. M. Stahr, Pesticide residues in clothing, Case study of a Midwestern farmer's coverall contamination, *J. Environ. Health 51*:273 (1989).
38. E. Kissa, Kinetics and mechanisms of soiling detergency, *Detergency Theory and Technology*, (W. G. Cutler and E. Kissa, eds.), Marcel Dekker, New York, 1987, pp. 193–332.
39. E. Kissa, Evaluation of detergency, *Detergency Theory and Technology* (W. G. Cutler and E. Kissa, eds.), Marcel Dekker, New York, 1987, pp. 1–90.
40. J. Laughlin and R. E. Gold, *Laundering Pesticide Contaminated Clothing*, NebGuide HEG 81-152, 1989. Lincoln, Nebraska: University of Nebraska Cooperative Extension.
41. T. H. Lillie, J. M. Livingston, and M. A. Hamilton, Recommendations for selecting and decontaminating pesticide applicator clothing, *Bull. Environ. Contamin. Toxicol. 27*:716 (1981).
42. T. H. Lillie, R. E. Hampson, Y. A. Nishioka, and M. A. Hamilton, Effectiveness of detergent and detergent plus bleach for decontaminating pesticide applicator clothing, *Bull. Environ. Contamin. Toxicol. 29*:89 (1982).
43. J. L. Keaschall, J. M. Laughlin, and R. E. Gold, Effect of laundering procedures and functional finishes on removal of insecticides selected from three chemical classes, *Performance of Protective Clothing*, STP #900 (R. L. Barker and G. C. Coletta, eds.), ASTM, Philadelphia, 1986, p. 162.
44. C. Nelson, J. Laughlin, C. Kim, K. Rigakis, M. Raheel, and L. Scholten, *Arch. Environ. Contamin. Toxicol. 23*(6):85 (1992).
45. C. B. Easley, J. M. Laughlin, R. E. Gold, and K. Schmidt, Detergents and water temperature as factors in methyl parathion removal from denim fabrics, *Bull. Environ. Contamin. Toxicol. 28*:239 (1982).
46. E. Easter, Removal of pesticide residues from fabrics by laundering, *Text. Chem. Color. 15*:29 (1983).
47. J. R. Fleeker, C. Nelson, M. F. Wazir, and M. M. Olsen, Removal of insecticide formulations containing carbaryl (Sevin) from apparel fabrics, *Performance of Protective Clothing*, STP #989 (S. Z. Mansdorf, R. Sager, and A. P. Nielsen, eds.), ASTM, Philadelphia, 1988, p. 715.
48. J. M. Laughlin, C. B. Easley, and R. E. Gold, Methyl parathion residues in contaminated fabrics after laundering, *Dermal Exposure Related to Pesticide*

Use, American Chemical Society Symposium Series #273, p. 177, 1985. Washington, D.C.: ACS.

49. C. B. Easley, J. M. Laughlin, R. E. Gold, and D. R. Tupy, Laundering procedures for removal of 2,4-dichlorophenoxyacetic acid ester and amine from herbicide contaminated fabrics, *Arch. Environ. Contamin. Toxicol. 12*:71 (1983).
50. E. P. Easter and J. O. DeJonge, The efficacy of laundering captan and Guthion contaminated fabrics, *Arch. Environ. Contamin. Toxicol. 14*:281 (1985).
51. M. Raheel, Laundering variables in removing carbaryl and atrazine residues from contaminated fabrics, *Bull. Environ. Contamin. Toxicol. 39*:671 (1987).
52. J. Laughlin, K. Newburn, and R. E. Gold, Pyrethroid insecticides and formulation as factors in residues remaining in apparel fabrics after laundering, *Bull. Environ. Contamin. Toxicol. 47*:355 (1991).
53. E. Braun, R. Frank, and G. M. Ritcey, Removal of organophosphorus, organochlorine and synthetic pyrethroid insecticides and organochlorine fungicides from coverall fabric laundering, *Bull. Environ. Contamin. Toxicol. 44*:92 (1990).
54. C. N. Nelson and J. R. Fleeker, eds., *Limiting Pesticide Exposure Through Textile Cleaning Procedures*, North Central Cooperative Series Bulletin #314, (1988). Fargo, N. D.: North Dakota State University.
55. J. H. Chiao-Cheng, B. M. Reagan, R. R. Bresee, C. E. Meloan, and A. M. Kadoum, Carbamate insecticide removal in laundering from cotton and polyester fabrics. *Arch. Environ. Contamin. Toxicol. 17*:87 (1988).
56. C. N. Nelson, J. R. Fleeker, and C. Janecek, Pesticide removal from clothing by laundering. *N. D. Farm Research Bull. 47*:23 (1989).
57. J. M Laughlin, J. L. Lamplot, and R. E. Gold, Chlorypyrifos residues in protective apparel fabrics following commercial or consumer refurbishment, *Performance of Protective Clothing*, ASTM STP #989 (S. Z. Mansdorf, R. Sager, and A. P. Nielsen, eds.), ASTM, Philadelphia, 1988, p. 705.
58. C. B. Easley, J. M. Laughlin, R. E. Gold, and R. M. Hill, Laundry factors influencing methyl parathion removal from contaminated denim fabric. *Bull. Environ. Contamin. Toxicol. 29*:461 (1982).
59. M. J. Olsen, C. Janecek, and J. R. Fleeker, Removal of paraquat from contaminated fabrics. *Bull. Environ. Contamin. Toxicol. 37*:558 (1986).
60. C. Goodman, J. Laughlin, and R. E. Gold, Strategies for laundering protective apparel fabric sequentially contaminated with methyl parathion, *Performance of Protective Clothing*, STP #989 (S. Z. Mansdorf, R. Sager, and A. P. Nielsen, eds.), ASTM, Philadelphia, 1988, p. 671.
61. A. Davidsohn and B. M. Milwidsky, *Synthetic Detergents*, John Wiley & Sons, New York, 1978.
62. H. von Rybinski, M. J. Schwuger, *Nonionic Surfactants Physical Chemistry* (M. J. Schick, ed.), Marcel Dekker, New York, 1987, pp. 45–108.
63. H. Lange and P. Jeschke, *Ionic Surfactants Physical Chemistry* (M. J. Schick, ed.), Marcel Dekker, New York, 1987, pp. 1–44.

64. M. Raheel and E. C. Gitz, Effect of fabric geometry on resistance to pesticide penetration and degradation, *Arch. Environ. Contamin. Toxicol. 14*:273 (1985).
65. J. Laughlin and R. E. Gold, Methyl parathion redeposition during laundering functionally finished protective apparel fabrics, *Bull. Environ. Contamin. Toxicol. 42*:691 (1989).
66. J. Parks, D. H. Branson and S. Burks, Pesticide decontamination from fabric by laundering and simulated weathering. *J. Environ. Sci. Health B25*:281 (1990).
67. C. J. Kim, J. F. Stone, and C. E. Sizer, Removal of pesticide residues as affected by laundering variables, *Bull. Environ. Contamin. Toxicol. 29*:95 (1982).
68. J. R. Fleeker, C. Nelson, A. W. Braaten, and J. B. Fleeker, Quantitation of pesticides on apparel fabrics, *Performance of Protective Clothing* STP #989 (S. Z. Mansdorf, R. Sager, and A. P. Nielsen, eds.), ASTM, Philadelphia, 1988, p. 745.
69. J. Laughlin and R. E. Gold, Methyl parathion residue retained in fabrics for functional clothing resulting from use of cationic fabric softeners in laundering, *Bull. Environ. Contamin. Toxicol. 44*:737 (1990).
70. D. N. Hild, J. Laughlin and R. E. Gold, Laundry parameters as factors in lowering methyl parathion residue in cotton/polyester fabrics, *Arch. Environ. Contamin. Toxicol. 18*:908 (1989).
71. K. B. Rigakis, S. Martin-Scott, E. M. Crown, N. Kerr, and B. Eggertson, Limiting pesticide exposure through textile cleaning procedures and selection of clothing, *Agri. Forest. Bull. 10*:24 (1987).
72. J. Laughlin and R. E. Gold, Levels of water hardness, detergent type and prewash product use as factors affecting methyl parathion removal, *Clo. Text. Res. J. 8*:61 (1990).
73. C. A. Popelka, *Comparison of Induction-Heat Extraction Method and Solvent Extraction Method for the Analysis of Pesticide Residues in Fabric,* master's thesis, Iowa State University, Ames, 1985.
74. J. Laughlin, Thermal and chemical energy in laundering to maximize methyl parathion removal from cotton/polyester fabrics, *Arch. Environ. Contamin. Toxicol. 20*:49 (1991).
75. R. M. Sagan and S. K. Obendorf, Starch as a non-durable finish to reduce the retention of methyl parathion on cotton and polyester fabrics after laundering, *Proceedings of the 1st International Symposium on the Impact of Pesticides, Industrial and Consumer Chemicals on the Environment*, 1988, p. 192. Athens, Georgia: University of Georgia.
76. C. B. Brown, S. H. Thompson, G. Stewart, Oil take-up and removal by washing, *Text. Res. J. 38*:735 (1968).
77. T. Fort, H. R. Billica, and T. H. Grindstaff. Studies of soiling and detergency. *Text. Res. J. 36*:99 (1966).
78. S. K. Obendorf and C. M. Solbrig, Distribution of the organophosphorus pesticides malathion and methyl parathion on cotton/polyester fabrics after

laundering as determined by electron microscopy, *Performance of Protective Clothing, ASTM* STP #900 (R. L. Barker and G. C. Coletta, eds.), pp. 187–204, ASTM, Philadelphia, 1986.

79. C. M. Solbrig and S. K. Obendorf, Distribution of residual pesticide within textile structures as determined by electron microscopy. *Text. Res. J. 55*:540 (1985).
80. C. B. Easley, J. M. Laughlin, and R. E. Gold, Methyl parathion removal from denim fabrics by selected laundry procedures. *Bull. Environ. Contamin. Toxicol. 27*:101 (1981).
81. J. M. Laughlin, C. B. Easley, R. E. Gold, and R. M. Hill, Fabric parameters and pesticide characteristics that impact on dermal exposure of applicators, *Performance of Protective Clothing*, STP #900 (R. L. Barker and G. C. Coletta, eds.), ASTM, Philadelphia, p. 136-150 (1986).
82. J. M. Laughlin, C. B. Easley, R. E. Gold, and D. R. Tupy, Methyl parathion transfer from contaminated fabrics to subsequent laundry and to laundry equipment, *Bull. Environ. Contamin. Toxicol. 27*:518 (1981).
83. J. Laughlin and R. E. Gold, Methyl parathion residues in functionally finished cotton and polyester after laundering and abrasion, *Clo. Text. Res. J. 5*:10 (1987).
84. J. Laughlin and R. E. Gold, Methyl parathion redeposition during laundering functionally finished protective apparel fabrics, *Bull. Environ. Contamin. Toxicol. 42*:691 (1989).
85. C. J. Kim, J. F. Stone, J. R. Coats, and S. J. Kadolph, Removal of alachlor residues from contaminated clothing fabrics, *Bull. Environ. Contamin. Toxicol. 36*:234 (1986).
86. E. F. R. Uyenco and S. K. Obendorf, Effect of functional finishes on retention of the pesticide malathion on polyester/cotton fabrics, *AATCC Book of Papers*, pp. 239–247 (1984).
87. N. E. Breen, D. J. Durnam, and S. K. Obendorf, Residual oily soil distribution on polyester/cotton fabric after laundering with selected detergents at various wash temperatures, *Text. Res. J. 54*:198 (1984).
88. J. J. Webb and S. K. Obendorf, Detergency study: Comparison of the distribution of natural residual soils after laundering with a variety of detergent products, *Text. Res. J. 57*:640 (1987).
89. S. A. Weglinski and S. K. Obendorf, Soil distribution on fabric after laundering. *Text. Chem. Color. 17*:21 (1985).
90. S. K. Obendorf, Electron microscopical study of soiling and soil removal. *Text. Chem. Color. 20*:11 (1988).
91. S. K. Obendorf and N. A. Klemash, Electron microscopical analysis of oily soil penetration into cotton and polyester/cotton fabrics, *Text. Res. J. 52*:434 (1982).
92. S. K. Obendorf, Y. M. N. Namasté, and D. J. Durnam, A microscopical study of residual oily soil distribution on fabrics of varying fiber content, *Text. Res. J. 53*:375 (1983).
93. C. J. Kim and J. Kim, Dispersion mechanism of an industrial pesticide

chemical in woven fabric structures, *Performance of Protective Clothing*, STP #989 (S. Z. Mansdorf, R. Sager, and A. P. Nielsen, eds.), ASTM, Philadelphia, 1988, pp. 680–691.

94. V. H. Freed, J. E. Davies, L. J. Peters, and F. Parveen, Minimizing occupational exposure to pesticides: repellency and penetrability of treated textiles to pesticide sprays. *Residue Review 75*:159 (1980).
95. J. Orlando, D. Branson, G. Ayers, and R. Leavitt, The penetration of formulated Guthion spray through selected fabrics. *J. Environ. Sci. Health 5*:617 (1981).
96. T. K. Das and A. K. Kulshreshtha, Soil release finishing of textiles. *J. Sci. Ind. Res. 38*:611 (1979).
97. K. P. Ringenberg, J. M. Laughlin, and R. E. Gold, Chlorpyrifos residue removal from protective apparel through solvent-based refurbishment procedures, *Performance of Protective Clothing*, STP #989, (S. Z. Mansdorf, R. Sager, and A. P. Nielsen, eds.), ASTM, Philadelphia, 1988, pp. 697–704.
98. J. Laughlin and R. E. Gold, Refurbishment of nonwoven protective apparel fabrics contaminated with methyl parathion, *Bull. Environ. Contamin. Toxicol. 45*:452 (1990).
99. D. H. Branson, J. O. DeJonge, and D. Munson, Thermal response associated with prototype pesticide protective clothing, *Text. Res. J. 56*:27 (1986).
100. D. Branson and S. Rajadhyaksha, Distribution of malathion on Gore-Tex® fabric before and after sunlight exposure and laundering as determined by electron microscopy, *Performance of Protective Clothing*, ASTM STP #989 (S. Z. Mansdorf, R. Sager, and A. P. Nielsen, eds.), ASTM, Philadelphia, 1988, pp. 651–659.
101. M. A. Huisman and M. A. Morris, Oily and particulate soils in fabrics after laundering, *Text. Res. J. 41*:657 (1971).
102. J. Laughlin and R. E. Gold, Methyl parathion residue removal from protective apparel fabrics soiled with synthetic sebum or vegetable oil, *Arch. Environ. Contamin. Toxicol. 19*:205 (1990).
103. J. Laughlin, Methyl parathion residues in protective apparel fabrics: Effect of residual soils on decontamination, *Performance of Protective Clothing IV*, ASTM STP #1133 (J. P. McBriarty and N. W. Henry, eds.). ASTM, Philadelphia, 1992.
104. J. M. Laughlin and R. E. Gold, The vaporization of methyl parathion from contaminated cotton fabrics. *Text. Chem. Color. 19*:39 (1987).
105. J. Laughlin and R. E. Gold, Evaporative dissipation of methyl parathion from laundered protective apparel fabrics, *Bull. Environ. Contamin. Toxicol. 42*:566 (1989).
106. C. J. Kim, S. J. Kadolph, and J. F. Stone, Effects of pretreatment, detergent, water hardness, drying method, and fiber content on fonofos residue removal from clothing fabrics, *Proceedings of the 1st International Symposium of Impact of Pesticide, Industrial, and Consumer Chemicals*, 1988, pp. 202–210. Athens, Georgia: University of Georgia.

107. C. J. Kim, Effects of convection-oven and microwave-oven drying on removal of alachlor residues in a fabric structure, *Bull. Environ. Contamin. Toxicol. 43*:904 (1989).

108. J. Laughlin, Textiles research and pesticides: A human resources perspective, *Human Resources Research 1887–1987* (R. E. Deacon and W. E. Huffman, eds.), Iowa State University, Ames, Iowa, 1986, pp. 61–74.

6

Heat Transfer Through Human Body and Clothing Systems

KENNETH C. PARSONS Loughborough University of Technology, Loughborough, Leicestershire, United Kingdom

I. INTRODUCTION

Successful protective clothing must allow the functions of the body to be maintained and account for its responses as well as protect it from environmental hazards and agents. Clothing provides a microclimate between the body and the external environment. The nude body exists within and responds to this microclimate, and the thermoregulatory responses of the body and the heat transfer and vapor permeation properties of the clothing determine the microclimate.

To provide for thermal comfort and health, protective clothing should maintain an internal body temperature within acceptable limits (e.g., 36°C–38°C) and allow skin temperature and skin wettedness to be within comfort limits. That internal body temperature should be relatively constant at around 37°C implies that heat production and any heat transfer into the body must be balanced by heat loss from the body (including that through clothing). A rational analysis of heat production within the body, its transfer to the skin surface, and heat exchange between skin, clothing, boundary air layer, and environment provides a fundamental theoretical and practical approach, which can contribute to the design and evaluation of clothing.

The dynamic heat production and exchange mechanisms between a clothed person and the environment are not fully understood. The approach taken has been to provide simple models of the system, particularly in terms of the properties and behavior of clothing. These models range from consideration of only the dry heat transfer through clothing, to more complex models involving vapor and moisture transport, pumping effects, breathability, ventilation, and interactions with the wearer's activity. Methods of representing the properties of clothing in terms of simple models are described here. Methods of determining the properties and an example of how they are used in the human heat balance equation are also presented.

II. A SIMPLE CLOTHING MODEL

The dry thermal insulation value of clothing materials and clothing ensembles is of fundamental importance and has been extensively investigated. A simple thermal model is of a heated body with a single layer of insulation in perfect contact with the skin (Figure 1). For the body to maintain equilibrium, heat flows to the skin, determining skin temperature, through the insulation to the clothing surface, determining clothing temperature, and to the outside environment. If the body were not continuously heated (i.e., death), heat would flow out of the body until equilibrium was reached when body temperature, skin temperature, and clothing temperature were at environmental temperature. For a continuously heated body (by metabolic heat production), a dynamic equilibrium is maintained where (generally) body temperature is greater than skin temperature, which is greater than temperature of the clothing surface, which in turn is greater than environmental temperature.

That temperature of clothing is higher than environmental temperature emphasizes that the environment also provides insulation (the boundary or air layer). The properties of this ''layer'' are very important to heat exchange and can be affected by the external environment. The basic or intrinsic clothing insulation I_{cl} will, according to the model, be independent of the external environmental conditions. The thermal properties of each of the components of the model presented in Figure 1 are discussed in the following sections.

A. Heat Transfer from the Body to the Skin

Metabolic heat production occurs in all parts of the body, and the thermoregulatory system regulates how much is transferred to the skin. This will involve heat transfer through tissues and will be greatly determined by the degree of vasodilation. Burton and Edholm [1] provide a detailed description. In terms of the model shown in Figure 1, it is important to know that this will affect skin condition (temperature and sweat), which is important for considering heat transfer through clothing.

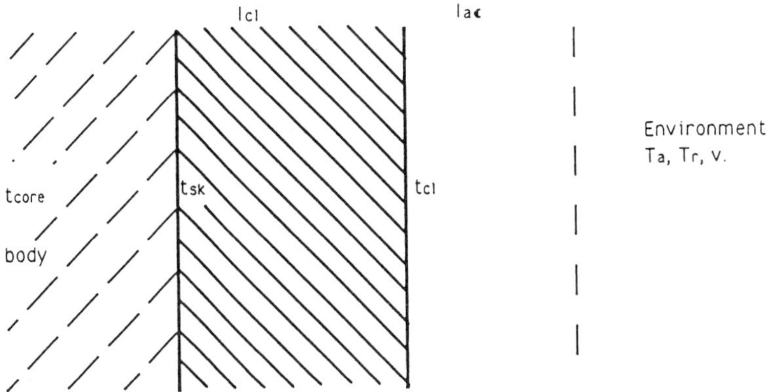

Figure 1 A simple model of heat transfer between the clothed body and the external environment. (From Ref. 16.)

B. Intrinsic Clothing Insulation (I_{cl})

Intrinsic (or basic) clothing insulation is a property of the clothing itself (and not the external environment or body condition) and represents the resistance to heat transfer between the skin and the clothing surface. Rate of heat transfer through the clothing is by conduction, which depends on surface area (m^2), temperature gradient between skin and clothing surface (°C), and the thermal conductivity of the clothing (W/m^2 °C). Intrinsic clothing insulation is the reciprocal of clothing conductivity with units of m^2 °C/W. Gagge et al. [2] first proposed the Clo unit. This was to replace the rather physical unit of m^2 °C/W with something easily visualized and related to clothing worn on the human body. One Clo was said to be the thermal insulation required to keep a sedentary person comfortable at 21°C. It is said to have an average value of 0.155 m^2 °C/W and is representative of the insulation of a typical business suit.

It is important to note that the m^2 term in this unit refers to surface area of the body. A neck tie worn on its own for example may be given an estimated thermal insulation value of 0.1 Clo. A suit made of the same material may have an estimated insulation value of 1.0 Clo. The important point, therefore, is that the Clo value gives an estimate of insulation as if any clothing was distributed evenly over the whole body. This can cause confusion for those used to considering the thermal insulation of materials. A unit of the thermal insulation of material is the Tog [3]. One Tog is given a value of 0.1 m^2 °C/W where the m^2 term refers to the area of the material tested.

C. Thermal Resistance of the Environment—The Air Layer (I_a)

If the environment had perfect conductivity (no resistance), the surface temperature of the clothing would fall to that of the environment. This is not the case, however. The environment provides significant thermal resistance depending upon how environmental conditions affect heat transfer (mainly by convection C and radiation R). The thermal resistance (insulation) of the environment for a nude body is

$$I_a = 1/h$$

where

$$h = h_r + h_c$$
h_r = radiative heat transfer coefficient
h_c = convective heat transfer coefficient

For a clothed body, the surface area for heat transfer is increased by an amount depending upon the thickness of the clothing layer. This is taken into consideration using the term f_{cl}, which is the ratio of the clothed surface area of the body to the nude surface area of the body. That is,

$$I_{ac} = 1/f_{cl}h$$

A difficulty is in determining a value for f_{cl}. On an active person, only an approximation can be made, and even on a copper manikin, it is difficult to measure [4]. More sophisticated systems include photographic techniques [5] and computer-aided anthropometric scanners [6]. An approximation can be based on the intrinsic clothing insulation value. For example, McCullough and Jones [4] give the following equation as a rough estimate for indoor ensembles:

$$f_{cl} = 1.0 + 0.31\ I_{cl} \qquad (I_{cl} \text{ in Clo})$$

It should be emphasized that this equation is an approximation, and other values of the coefficient for I_{cl} have been used (e.g., 0.15 by Fanger [7]).

D. Total Insulation (I_T) and Effective Insulation (I_{cle})

The necessary principle and quantities for the model shown in Figure 1 are provided. There are, however, other values, which are mentioned in the scientific literature, and it is useful to briefly discuss these here. Total clothing insulation is the combined insulation provided by clothing and surrounding air layer. Parsons [8] gives the following equation:

$$I_T = I_{cl} + I_{ac}$$

where I_{ac} is the air layer around the clothed body. That is,

$$I_T = I_{cl} + I_a/f_{cl} = I_{cl} + 1/f_{cl}h$$

To determine I_{cl}, by using a thermal manikin for example, it is possible to measure I_T using a clothed manikin and I_a using a nude manikin. Determination of f_{cl} is, however, less accurate. From the preceding equation

$$I_{cl} = I_T - I_a/f_{cl}$$

A more convenient term for measurement is, therefore, effective clothing insulation I_{cle}, where

$$I_{cle} = I_T - I_a$$

Although more convenient to determine, I_{cle} is not consistent with the model provided in Figure 1 and can lead to confusion.

E. The Burton Thermal Efficiency Factor (F_{cl})

Oohori et al. [9] provide the following equation for the dry heat loss from the skin:

$$\text{Dry} = (F_{cl}f_{cl}h)\,(T_{sk} - T_a)$$

where

F_{cl} is the Burton thermal efficiency factor
$F_{cl} = I_{ac}/(I_{cl} + I_{ac})$
$I_{ac} = 1/f_{cl}h$ as before

Efficiency factors (clothing indices, etc.) are often useful for simplifying calculations, providing terms that have physical meaning, and linking with experimental method. They can, however, cause confusion, and the essential concepts for the model provided in Figure 1 are I_{cl} and I_{ac}. The main point about this model is that I_{cl} is intrinsic to the clothing and is not affected by external conditions. That the model may be inadequate in many conditions will be discussed later. Methods of determining I_{cl} values are discussed next.

F. Method of Determining the Dry Thermal Insulation of Clothing

The thermal insulation of clothing materials can be measured on standardized equipment, which usually involves placing a sample of material on the equipment; by measuring heat flows or temperature, the thermal insulation can be calculated. Such equipment would include standardized heated flat plates and cylinders. More sophisticated methods involve a heated copper man (manikins), with a temperature distribution across the body similar to that of a human subject (see McCullough and Jones [4], Kerslake [10], Oleson et al. [5], Wyon et al. [11] for reviews).

Table 1 Dry Clothing Insulation (I_{cl}) Values for Whole Clothing Ensembles

Work clothing	I_{cl} Clo	I_{cl} m² °C/W
Underpants, boiler suit, socks, shoes	0.70	0.110
Underpants, shirt, trouser, socks, shoes	0.75	0.115
Underpants, shirt, boiler suit, socks, shoes	0.80	0.125
Underpants, shirt, trousers, jacket, socks, shoes	0.85	0.135
Underpants, shirt, trousers, smock, socks, shoes	0.90	0.140
Underwear with short sleeves and legs, shirt, trousers, jacket, socks, shoes	1.00	0.155
Underwear with short legs and sleeves, shirt, trousers, socks, shoes	1.10	0.170
Underwear with long legs and sleeves, thermojacket, socks, shoes	1.20	0.185
Underwear with short sleeves and legs, shirt, trousers, jacket, thermojacket, socks, shoes	1.25	0.190
Underwear with short sleeves and legs, boiler suit, thermojacket and trousers, sock, shoes	1.40	0.220
Underwear with short sleeves and legs, shirt, trousers, jacket, thermojacket and trousers, sock, shoes	1.55	0.225
Underwear with short sleeves and legs, shirt, trousers, jacket, heavy quilted outer jacket and overalls, socks, shoes	1.85	0.285
Underwear with short sleeves and legs, shirt, trousers, jacket, heavy quilted outer jacket and overalls, socks, shoes, cap, gloves	2.00	0.310
Underwear with long sleeves and legs, thermojacket and trousers, outer thermojacket and trousers, socks, shoes	2.20	0.340
Underwear with long sleeves and legs, thermojacket and trousers, parca with heavy quilting, overalls with heavy quilting, socks, shoes, cap, gloves	2.55	0.395

Source: Ref. 13.

Thermal manikins are specialized and expensive and are not widely available. Dry thermal insulation values have been determined for many types of clothing using thermal manikins, and databases (tables) of insulation values have been provided. ISO 9920 [12] provides such a database as do McCullough and Jones [4]. Oleson and Dukes-Dubos [13] provide examples for whole clothing ensembles (Table 1) and for individual clothing garments (Table 2).

Values of I_{cl} are obtained directly from Table 1. However it is likely in practice that the clothing ensembles in the database (tables) will not be identical to the ensemble for which an I_{cl} value is required. In this case, estimates of I_{cl} values can be made from I_{clu} values which are "effective" insulation values for garments, that is, insulation values calculated without taking account of an increase in surface area due to the garment. ISO 9920 [12] suggests that simply adding effective Clo values for garments gives a realistic estimate of ensemble I_{cl} values. This was different from other methods where simple regression equations were used to estimate the insulation of clothing ensembles from garments values (e.g., [14]).

ISO 9920 [12] provides tables of insulation values for clothing ensembles and garments. Garments can be identified in terms of garment type (e.g., shirt), style (e.g., long sleeved, short sleeved), and material type (e.g., 100% cotton, cotton polyester). Parker and Parsons [15] describe a computer database of the ISO clothing ensembles and a method of matching required clothing with those for which I_{cl} values are required. This is a most effective way to use such a database. McCullough and Jones [4] report on a user trial of their database "manual" and found that even clothing science students were unable to use it effectively.

ISO 9920 [12] provides the following equations for estimating the I_{clu} for garments and, hence, I_{cl} values for clothing ensemble:

$$I_{clu} = 0.095 \times 10^{-2} A_{cov} \quad (\mathrm{m^2\ ^\circ C\ W^{-1}})$$

$$I_{cl} = \Sigma I_{clu,i} \quad (\mathrm{m^2\ ^\circ C\ W^{-1}})$$

where

I_{cl} = intrinsic clothing insulation for clothing ensemble
$I_{clu,i}$ = effective insulation for garment i
A_{cov} = body surface area covered by the garment (%)

The model in Figure 1 provides a representation of clothing in the stationary, comfortable or cold, human body in many conditions. It provides only an approximation, however, to many other circumstances.

Table 2 Effective Dry Clothing Insulation (I_{clu}) Values for Garments

Garment description	Thermal insulation (I_{clu}), Clo	Garment description	Thermal insulation (I_{clu}), Clo
Underwear		Thin sweater	0.20
Panties	0.03	Sweater	0.28
Underpants with long legs	0.10	Thick sweater	0.35
Singlet	0.04	Jackets	
T-shirt	0.09	Light summer jacket	0.25
Shirt with long sleeves	0.12	Jacket	0.35
Panties + bra	0.03	Smock	0.30
Shirts, Blouses		High insulative, fiber-pelt	
Short sleeves	0.15	Boilet suit	0.90
Light weight, long sleeves	0.20	Trousers	0.35
Normal, long sleeves	0.25	Jacket	0.40
Flannel shirt, long sleeves	0.30	Vest	0.20
Light weight blouse, long sleeves	0.15	Outdoor clothing	
Trousers		Coat	0.60
Shorts	0.06	Down jacket	0.55
Light weight	0.20	Parca	0.70
Normal	0.25	Fiber-pelt overalls	0.55
Flannel	0.28	Sundries	
Dress, skirts		Socks	0.02
Light skirt (summer)	0.15	Thick ankle socks	0.05
Heavy dress (winter)	0.25	Thick long socks	0.10
Light dress, short sleeves	0.20	Nylon stockings	0.03
Winter dress, long sleeves	0.40	Shoes (thin soled)	0.02
Boiler suit	0.55	Shoes (thick soled)	0.04
Sweaters		Boots	0.10
Sleeveless vest	0.12	Gloves	0.05

Source: Ref. 13.

III. THE TWO-PARAMETER MODEL

An important limitation of the model shown in Figure 1 is that it does not consider wet clothing. Moisture can transfer heat between the body and the environment. This is particularly important when the skin sweats. A simple two-parameter model of clothing would be to consider dry heat transfer (Figure 1) and moisture transfer (Figure 2) as separate and independent mechanisms that combine to provide the total effect (Figure 3).

A. Intrinsic Resistance of Clothing to Vapor Transfer (I_{ecl})

The model in Figure 2 is analogous to that of Figure 1. In Figure 1, however, the temperature gradient between the skin and the outside surface of the clothing provides the driving potential for heat loss. In Figure 2, it is the vapor pressure difference between skin and environment that provide the driving potential. Liquid (sweat) on the skin evaporates at the skin surface and is transported through the clothing to the environment. The resistance to this vapor transfer is termed the *intrinsic evaporative resistance* I_{ecl}. The units of I_{ecl} are, therefore, m^2kPa/W. The partial vapor pressure at the skin is assumed to be the saturated vapor pressure at skin temperature. From Antoine's equation,

$$P_{sk,s} = \exp\left(18.956 - \frac{4030}{t_{sk} + 235}\right) \text{ mb}$$

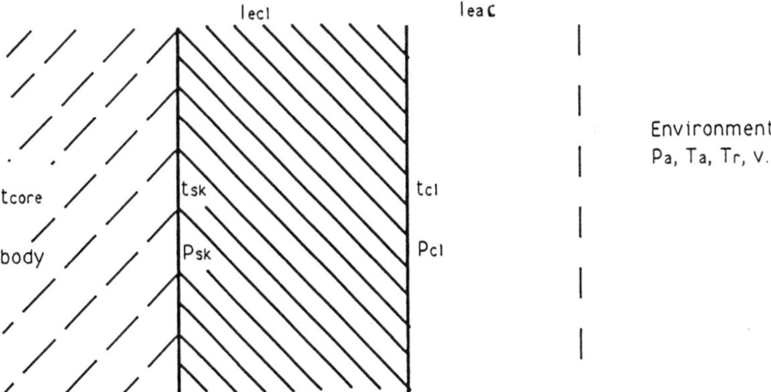

Figure 2 A model of vapor transfer between the clothed body and the external environment. (From Ref. 16.)

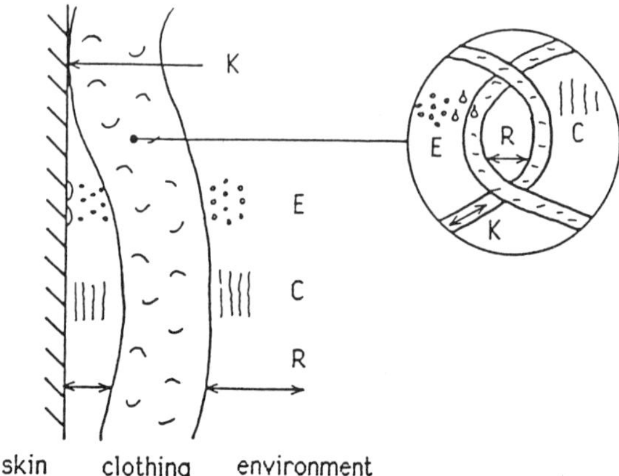

Figure 3 Simple two parameter description of clothing. (From Ref. 17.)

The partial vapor pressure in the air (P_a) is related to relative humidity (ϕ) by

$$\phi = P_a/P_{sa}$$

where P_{sa} is the saturated vapor pressure in air at air temperature and can be obtained by substituting P_{sa} and T_a for $P_{sk,s}$ and T_{sk}, respectively, in Antoine's equation.

B. Resistance of the Environment to Vapor Transfer from Clothing (I_{ea})

When vapor reaches the clothing surface, it transfers to the environment. This will depend upon the evaporative heat transfer coefficient h_e. It has been traditional to determine this value by relating it to the convective heat transfer coefficient by the Lewis number.

The Lewis number is defined as the ratio of mass transfer coefficient by evaporation to heat transfer coefficient by convection (no radiation). At sea level for air,

$$L_R = h_e/h_c = 2.2 \qquad (\text{kTorr}^{-1})$$

The Lewis relationship L_R is not affected by the size and shape of the body, nor by air speed or temperature. It is affected by the physical properties of the gases involved (in this case, of air and water vapor) and atmospheric pressure. For air,

$$L_R = 1.65/P_{atm} \qquad (\text{K/mb})$$

where

$$P_{atm} = \text{atmospheric pressure in atmospheres [18]}$$

The maximum evaporation from the skin surface to air is, therefore,

$$E_{max} = h_e(P_{sk,s} - P_a) \qquad (\mathrm{W/m^2})$$

where

$$h_e = 1.65\ h_c\ \mathrm{W/m^2mb} = 16.5\ h_c\ \mathrm{W/m^2kPa}$$

In terms of the model in Figure 2,

$$h_e = 1/I_{eac} \cdot f_{cl}$$

or

$$I_{eac} = 1/h_e \cdot f_{cl} = 1/16.5\ h_c \cdot f_{cl}$$

where I_{ea} is the thermal resistance of the air layer. The maximum heat loss from the skin, through clothing, to the environment is, therefore,

$$E_{max} = \frac{1}{I_{eac} + I_{ecl}} (P_{s,sk} - P_a) \qquad (\mathrm{W/m^2})$$

For the case where the skin is not completely wet the skin wettedness w is used to calculate evaporative heat loss E where

$$w = E/E_{max}$$

that is

$$E = \frac{w}{I_{eac} + I_{ecl}} (P_{s,sk} - P_a)$$

With respect to vapor transfer through clothing, therefore, vapor transfer resistance of the air layer is

$$I_{eac} = 1/h_e \cdot f_{cl} = 1/16.5\ h_c \cdot f_{cl}$$

Intrinsic vapor transfer resistance of clothing I_{ecl} is, therefore, required. An important point is that the human skin is always "wet" to some extent. There is always a flow of vapor from wet tissues below the skin to drier air above (diffusion through skin − insensible heat loss). The vapor permeation properties of clothing are, therefore, of great importance.

C. Vapor Permeability Indices

To provide easy-to-understand values of clothing permeability or to aid in simplifying the heat transfer equation, a number of vapor permeability indices of clothing have been produced. Although all are essentially based on the two-parameter model, they have been developed independently and only recently has the relationship between each of them been determined [19]. As with the indices for dry clothing insulation, it is debatable whether the apparent advantage of simplification, provided by the indices, outweighs the ability to directly relate terms to the three clothing model parameters (Figure 2) and to refer directly to the fundamental units of physics.

D. Woodcock Moisture Permeability Index (i_m)

A property of clothing is that it impedes evaporative heat transfer more than it does sensible heat transfer [10]. This property can be expressed by the ratio of thermal conductivity for water vapor and sensible heat, respectively. In terms of Figures 1 and 2,

$$\frac{1/I_{ecl}}{1/I_{cl}} = \frac{I_{cl}}{I_{ecl}}$$

Woodcock [20] proposed the permeability index i_m, which compares this property with that for air (h_e/h_c). The permeability index of a material is

$$i_m = \frac{heT/hT}{h_e/h_c}$$

Woodcock [20] considered this as the ratio of the actual evaporative heat transfer, as hindered by clothing, to that of an aspirated wet bulb thermometer with the same dry heat transfer resistance. The equation for evaporative heat transfer then becomes

$$E = \frac{16.5\ i_m}{I_T}\ (P_{s,sk} - P_a)$$

The i_m value does not provide a value intrinsic to clothing but is affected by environmental conditions. Values of 0 (impermeable) to 1 (air) provide the range and are theoretically possible, but in practice typical values of around 0.5 are obtained for nude subjects, 0.4 for typical clothing, and 0.2 for impermeable-type clothing.

E. Permeation Efficiency Factor (F_{pcl})

Nishi and Gagge [21] proposed a permeation efficiency factor F_{pcl}, where

$$F_{pcl} = \frac{I_{ea}}{I_{ea} + I_{ecl}}$$

where

I_{ea} = the resistance of air to the transfer of water vapor
I_{ecl} = the resistance of clothing to the transfer of water vapor (i.e., analogous to Burton F_{cl} for dry heat exchange)

From the preceding equation,

$$F_{pcl} = \frac{1}{1 + I_{ecl}/I_{ea}}$$

From the Lewis relation,

$$I_{ea} = 1/h_e = 1/16.5\ h_c$$

Also for clothing we could consider a Lewis number K, where

$$I_{ecl} = 1/h_{ecl}$$
$$I_{cl} = 1/h_{cl}$$
$$K = h_{ecl}/h_{cl}$$

therefore,

$$1/h_{ecl} = I_{cl}/K$$

therefore,

$$F_{pcl} = \frac{1}{1 + (16.5/K)h_c\ \ I_{cl}}$$

Nishi [22] conducted empirical experiments involving light clothing and naphthalene sublimation and vapor transfer and relating it to water vapor transfer giving

$$F_{pcl} = \frac{1}{1 + 0.143\ h_c I_{clo}}$$

where

I_{clo} = effective clothing insulation in Clo

Further work by Lotens and Linde [23] and Oohori et al. [9] provided the current form of the equation:

$$F_{pcl} = \frac{1}{1 + 0.344 h_c I_{clo}}$$

It should be noted that I_{clo} is effective clothing insulation and, therefore, does not correct for the increase in surface area due to clothing.

Evaporative heat loss from a clothed body is then estimated as

$$E = F_{pcl} \cdot h_e (P_{sa} - P_a) \qquad (\mathrm{W/m^2})$$

Oohori et al. [9] describe the relationship between permeation ratios. They give

$$I_{ecl} = \frac{I_T}{16.5\ i_m} - \frac{I_{ea}}{f_{cl}}$$

The usefulness of clothing indices can be seen in the preceding equation. However, index values for clothing system are not widely available, and the index values do not directly demonstrate the mechanisms of the model. In addition, these values depend upon environmental conditions (according to the model), unlike the intrinsic values I_{cl} and I_{ecl}. Intrinsic clothing insulation values for dry heat transfer I_{cl} are now available for a wide variety of garments and clothing ensemble [4,12]. Values of I_{ecl} are required and some exist (Table 3). It could now be argued that the index values have "outlived their usefulness" and what are required for application are databases of intrinsic values for clothing.

F. Modification of the Two-Parameter Model—Wicking

For most practical applications, the simple dry insulation model is used to quantify clothing insulation (e.g., [7]). For more specialist evaluations of clothing, the two-parameter model is needed especially for the assessment of hot environments where sweating and, hence, vapor permeation properties will be of great importance.

For a more detailed representation of clothing, models are required that involve other important factors of clothing thermal behavior. A simple modification to the two-parameter model is to consider wicking of liquid through clothing materials with evaporation occurring within clothing. Kerslake [10] provides a model that is shown in Figure 4. The latent heat of vaporization is then removed from within the clothing and not at the skin.

McIntyre [18] summarizes Kerslake's work. The result is a loss of efficiency of sweating. In a simple case where the sweat is wicked through to the outer layer of clothing, this efficiency η becomes

$$\eta = \frac{1}{1 + 0.155hI_{clo}}$$

The efficiency term means that if the total evaporation rate is E_{sw}, then only ηE_{sw} is removed from the body. Evaporation does not always occur at the clothing surface, however, and in addition, condensation can supply heat to the clothing.

Table 3 Moisture Permeability Values for Typical Clothing Ensembles

Ensemble	Total evaporative resistance ($R_{e,t}$), kPa m^2/W	Moisture permeability index (i_m)	Evaporative resistance of clothing[a] ($R_{e,cl}$), kPa m^2/W	Moisture permeability index for clothing (i_{cl})
Men's business suit	0.044	0.37	0.033	0.32
Women's business suit	0.033	0.40	0.028	0.35
Men's summer casual	0.027	0.43	0.013	0.35
Jeans & shirt	0.031	0.40	0.020	0.32
Summer shorts & shirt	0.023	0.42	0.010	0.34
Women's casual	0.025	0.45	0.014	0.41
Women's shorts & tank top	0.022	0.40	0.009	0.27
Athletic sweat suit	0.029	0.43	0.017	0.41
Sleepwear & robe	0.035	0.41	0.024	0.37
Overalls & shirt	0.035	0.40	0.024	0.35
Insulated coverall & long underwear	0.042	0.39	0.037	0.35
Work shirt & trousers	0.037	0.40	0.025	0.34
Cleanroom coverall	0.039	0.38	0.023	0.32
Wool coverall	0.042	0.38	0.031	0.33
Firestop cotton coverall	0.038	0.40	0.027	0.38
Modacrylic coverall	0.038	0.41	0.027	0.36
Tyvak coverall	0.015	0.33	0.034	0.25
GoreTex 2-piece suit	0.044	0.39	0.033	0.33
Novex coverall	0.033	0.40	0.028	0.35
PVC/polyester knit acid suit	0.109	0.15	0.024	0.11
PVC/vinyl acid suit	0.125	0.10	0.115	0.09
Neoprene nylon suit	0.120	0.14	0.199	0.10

[a]Calculated based on an air layer resistance of 0014 kPa m^2/W.

Source: Ref. 24.

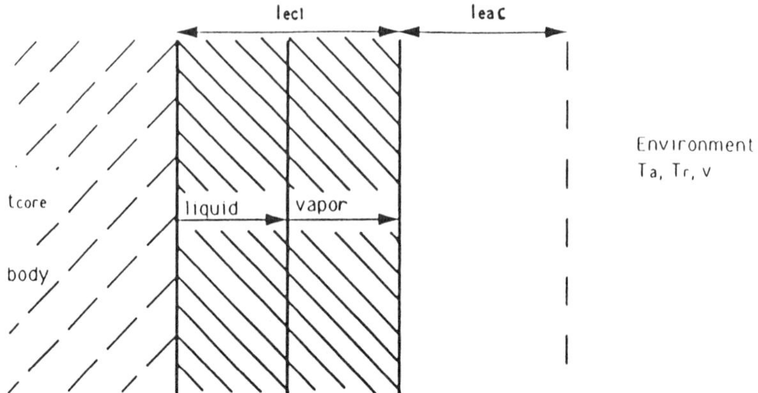

Figure 4 Model showing evaporation within clothing. (From Ref. 10.)

G. More Complex Clothing Models

Factors that have not been included in the models already described can have significant effects on the thermal properties of clothing. Kerslake [10] considers some of the practical effects on the thermal properties. He noted that insulation is provided by fabrics themselves and the layers of air trapped between the skin and clothing and the clothing layers. The insulation of fabrics is mainly due to the air trapped in and between them. On a clothed person, heat will be exchanged by penetration of air (or expulsion due to human movement) through vents and openings and directly through layers of the fabric. A simple model of this is provided in Figure 5.

The interaction between wind penetration, pumping, clothing ventilation, and thermal insulation of fabrics is not easily modeled and depends greatly on clothing design, thermal condition of the body, and a person's activity. Lotens [25] describes a four-layer model of clothing consisting of underclothing, trapped air, outer clothing, and adjacent air layer, including ventilation through apertures. Lotens states that the model was evaluated for the effects of (transient) moisture absorption, condensation, semipermeability, heat radiation, and ventilation. It is intended to link the model of clothing with a model of the human body and thermoregulation to provide a whole-body model of the physiological response of clothed humans.

More complex models of clothing, therefore, can have a number of characteristics in addition to the simple two-parameter approach. No fully comprehensive model exists, and the possibility for this is restricted. As inputs to such a model, a detailed description of the human skin (e.g.,

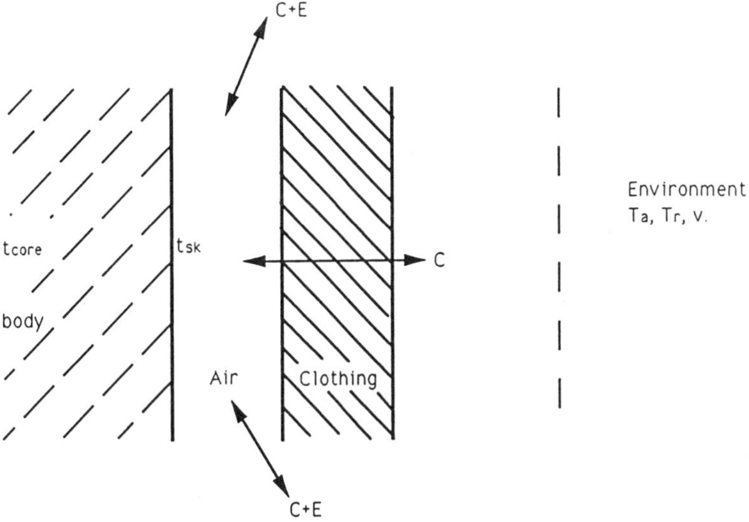

Figure 5 Air trapped between clothing and skin. (From Ref. 10.)

involving a thermal model of the human body) and the nature and properties of the clothing, environment, and behavior of the person would all be required. Even for relatively steady-state conditions, knowledge does not exist for a fully comprehensive model. For transient conditions where less is known about the effects of clothing and where thermal inertia (mass) of the clothing will be an important consideration, few models are available. Linking transient and steady-state responses in a truly comprehensive dynamic model of clothing (interfaced with body and environment) has not yet been achieved.

IV. THE HEAT BALANCE EQUATION FOR THE CLOTHED HUMAN BODY USING THE TWO-PARAMETER APPROACH

That the internal temperature should be maintained at around 37°C dictates that there is a heat balance between the body and its environment. That is, on average, heat transfer into the body and heat generation within the body must be balanced by heat outputs from the body. That is not to say that a steady state occurs. A steady state involves unchanging temperatures. Temperatures within the body and avenues of heat exchange will vary; the point is that for a constant temperature there will be a dynamic balance. If heat generation and inputs were greater than heat outputs, the body temperature would rise; if heat outputs were greater, the body temperature would fall. The heat balance equation for the human body can be represented in many forms. However, all equations have the same underlying concept and involve three types of terms: those for heat generation in the body, heat transfer, and heat storage.

The metabolic rate of the body M provides energy to enable the body to do mechanical work W and the remainder is released as heat (i.e., $M - W$). Heat transfer can be by conduction K, convection C, radiation R, and evaporation E. When combined together, all of the rates of heat production and loss provide a rate of heat storage S. For the body to be in heat balance (i.e., constant temperature), the rate of heat storage is zero ($S = 0$). If there is a net heat gain, storage will be positive, and body temperature will rise. If there is a net heat loss, storage will be negative, and body temperature will fall.

A. The Conceptual Equation

The conceptual heat balance equation is

$$M - W = E + R + C + K + S$$

i.e., for heat balance ($S = 0$)

$$(M - W) - E - R - C - K = 0$$

where

$(M - W)$ is always positive

E, R, C, and K are rates of heat loss from the body (i.e., positive value is heat loss, negative value is heat gain).

B. Units

It is important that all of these terms can be expressed as rates of heat production or loss. This allows a simple addition of gains and losses over each term. The units of the rate of energy gain or loss are energy per second, that is, joules per second ($J \cdot s^{-1}$) or watts (W). It is traditional (and useful) to "standardize" over persons of different sizes by using units of Watts per square meter of the total body surface area. The units are then W/m^2.

C. Body Surface Areas

Total body surface area is traditionally estimated from the simplified equation of Dubois and Dubois [26]; that is,

$$A_D = 0.202W^{0.425} \cdot H^{0.725}$$

where

$$A_D = \text{Dubois surface area (m}^2\text{)}$$
$$W = \text{weight of body (kg)}$$
$$H = \text{height of body (m)}$$

A standard value of 1.8 m^2 is sometimes used for a 70-kg man of height 1.73 m.

It is recognized that A_D provides only an estimate of body surface area. There are more accurate equations; however, A_D is traditionally used, and the nature of the heat balance equation means that any error is systematic and unimportant. In addition, objects of the same shape but different size have different heat transfer coefficients. This can be important but is not usually considered.

D. The Practical Heat Balance Equation

The principles and concept of the heat balance equation have already been described. For an analysis of heat exchange between the body and the environment and, hence, to quantify components of the equation to allow overall calculations (e.g., to make it useful in practice), there are two objectives:

1. The specific avenues of heat production and exchange for the human body must be identified.
2. Equations must be determined for the calculation of (estimate of) heat production and exchange. It is important for practical applications that terms in the equations are those that can be measured or estimated (e.g., the basic parameters of air temperature, mean radiant temperature, air velocity, relative humidity, activity, and clothing insulative properties).

ASHRAE [27] gives the following equation of heat balance:

$$M - W = Q_{sk} + Q_{res} = (C + R + E_{sk}) + (C_{res} + E_{res})$$

where

M = rate of metabolic energy production (W m^{-2})
W = rate of mechanical work (W m^{-2})
Q_{res} = total rate of heat loss through respiration (W m^{-2})
Q_{sk} = total rate of heat loss from the skin (W m^{-2})
C_{res} = rate of convective heat loss from respiration (W m^{-2})
E_{res} = rate of evaporative heat loss from respiration (W m^{-2})
C = rate of convective heat loss from the skin (W m^{-2})
R = rate of radiative heat loss from the skin (W m^{-2})

E_{sk} = rate of total evaporative heat loss from the skin where $E_{sk} = E_{rsw} + E_{dif}$ $(\mathrm{W \cdot m^{-2}})$

E_{rsw} = rate of evaporative heat loss from the skin through sweating $(\mathrm{W \cdot m^{-2}})$

E_{dif} = rate of evaporative heat loss from the skin through moisture diffusion $(\mathrm{W \cdot m^{-2}})$

A practical approach, is, therefore, to consider heat production within the body (M − W), heat loss at the skin ($C + R + E_{sk}$), and heat loss due to respiration ($C_{res} + E_{res}$). The next objective is to quantify components of the heat balance equation in terms of parameters that can be determined (measured or estimated).

E. Heat Production Within the Body ($M - W$)

Heat production within the body is related to the activity of the person. In general, oxygen is taken into the body (i.e., breathing air) and is transported by the blood to the cells where it is used to burn food. Most of the energy released is in terms of heat. Depending upon the activity, some external work will be performed. (Energy for mechanical work will vary from about zero (for many activities) to no more than 25% of total metabolic rate.

F. Heat Loss at the Skin ($C + R + E_{sk}$)

1. *Sensible Heat Loss* ($R + C$)

ASHRAE [27] gives the following derivation:

$$C = f_{cl}h_c\,(t_{cl} - t_a)$$
$$R = f_{cl}h_r\,(t_{cl} - t_r)$$
$$(C + R) = f_{cl}h\,(t_{cl} - t_o) \qquad (1)$$

where

$$t_0 = \frac{h_r t_r + h_c t_a}{h_r + h_c}$$

and

$$h = h_r + h_c$$

The actual transfer of heat through clothing (conduction, convection, and radiation) are combined into a single thermal resistance value R_{cl}, so

$$C + R = \frac{t_{sk} - t_{cl}}{R_{cl}} \qquad (2)$$

Combining Eqs. [1] and [2] to remove t_{cl}, the final form is

$$C + R = \frac{t_{sk} - t_o}{R_{cl} + 1/f_{cl} \cdot \mathrm{h}} \tag{3}$$

where C and R are as already defined and

f_{cl} = clothing area factor (the surface area of the clothed body A_{cl}, divided by the surface area of the nude body A_D.) (ND)
h_c = convective heat transfer coefficient ($\mathrm{Wm^{-2}K^{-1}}$)
h_r = linear radiative heat transfer coefficient ($\mathrm{Wm^{-2}K^{-1}}$)
t_o = operative temperature (°C)
t_r = mean radiant temperature (°C)
t_a = air temperature (°C)
h = combined heat transfer coefficient ($\mathrm{Wm^{-2}K^{-1}}$)
R_{cl} = thermal resistance of clothing ($\mathrm{m^2KW^{-1}}$)
t_{cl} = mean temperature over the clothed body (°C)
t_{sk} = mean skin temperature (°C)

Air temperature t_a, mean radiant temperature t_r, and the thermal resistance of clothing R_{cl} are all basic parameters that must be measured or estimated to define the environment. Mean skin temperature can be estimated as a constant value (e.g., around 33°C for comfort and 36°C under heat stress) or predicted from a dynamic model of human thermoregulation (see [27]). All other preceding values can be calculated from values of basic parameters. This is discussed later in the section on heat transfer. However, for a seated person, Mitchell [28] gives

$$h_c = 8.3v^{0.6} \quad \text{for} \quad 0.2 < v < 4.0$$

and

$$h_c = 3.1 \quad \text{for} \quad 0 < v < 0.2$$

where

v is the air velocity in $\mathrm{ms^{-1}}$ (i.e., a basic parameter)
The radiative heat transfer coefficient h_r can be given by

$$h_r = 4\epsilon \cdot \sigma(A_r/A_D)\ [273.2 + (t_{cl} + t_r)/2]^3$$

where

ϵ = the area weighted average emissivity of the clothing/body surface (ND)
σ = Stefan–Boltzmann constant, 5.67×10^{-8} ($\mathrm{Wm^{-2}K^{-1}}$)
A_r = effective radiative area of the body ($\mathrm{m^2}$)

ϵ is often assumed to be between 0.95 to 1.0. A_r/A_D can be estimated as 0.70 for a sitting person and 0.73 for a standing person [29]. t_r is a basic parameter, and t_{cl} must be calculated using iteration techniques. ASHRAE [27] suggests that a value of $h_r = 4.7\ \mathrm{Wm^{-2}K^{-1}}$ is a reasonable approximation for "most typical indoor conditions."

2. *Evaporative Heat Loss from the Skin* (E_{sk})

ASHRAE [27] gives the following equation:

$$E_{sk} = \frac{w(P_{sk,s} - P_a)}{R_{e,cl} + 1/(f_{cl}h_e)} \tag{4}$$

where

P_a = water vapor pressure in the ambient air (kPa)
$P_{sk,s}$ = water vapor pressure at the skin, normally assumed to be that of saturated water vapor at skin temperature, t_{sk} (kPa)
$R_{e,cl}$ = evaporative heat transfer resistance of the clothing surface ($\mathrm{m^2kPaW^{-1}}$)
h_e = convection evaporative heat transfer resistance at the clothing surface ($\mathrm{W/m^2kPa}$)
w = skin wettedness; the fraction of wetted skin area (ND)

h_e is calculated using the Lewis relation $h_e = LR \cdot h_c$. This is an important development in the establishment of the body heat balance equation allowing comparison and combination of dry and evaporative heat transfer.

The values P_a and $R_{e,cl}$ are basic parameters. $P_{sk,s}$ is calculated from Antoine's equation using the value for mean skin temperature t_{sk} (assumed constant value or from a dynamic model of human thermoregulation). Skin wettedness varies from a value of 0.06 when only natural diffusion of water through the skin occurs (E_{dif}) to 1.0 when skin is completely wet and maximum evaporation occurs (E_{max}). E_{max} can be calculated by letting $w = 1.0$ in Eq. (4). Skin wettedness can be calculated from

$$w = 0.06 + 0.94\,\frac{E_{rsw}}{E_{max}} \tag{5}$$

where

$$E_{rsw} = M_{rsw}h_{fg} \tag{6}$$

h_{fg} = heat of vaporization of water
= $2430\ \mathrm{kJ \cdot kg^{-1}}$ at 30°C

M_{rsw} = rate at which sweat is secreted $(\mathrm{kg \cdot s^{-1} \cdot m^{-2}})$

E_{max} can be determined from the basic parameters and Eq. (4). Skin evaporation and skin wettedness are often calculated from the heat balance equation in terms of wettedness required w_{req} or evaporation required E_{req} to maintain the body in heat balance. They can be calculated by determining E_{rsw} using Eq. [6] and

$$M_{rsw} = 4.7 \times 10^{-5}\ WSIG_b \exp(WSIG_{sk}/10.7)$$

where

$$WSIG_b = t_b - t_{b,n} \quad \text{for } t_b > t_{b,n}$$

$$WSIG_{sk} = t_{sk} - t_{sk,n} \quad \text{for } t_{sk} > t_{sk,n}$$

mean body temperature $t_b = \alpha t_{sk} + (1 - \alpha)\, t_{cr}$ (i.e., a weighted average of skin and core temperature— weighting α depending upon degree of vasodilation).

$t_{sk,n}$ = 33.7°C (i.e., mean skin temperature for a person in thermal neutrality

$t_{cr,n}$ = 36.8°C (i.e., core temperature in thermal neutrality)

Typical values for α are 0.2 for thermal equilibrium while sedentary, 0.1 in vasodilation, and 0.33 for vasoconstriction.

The mean body temperature in neutrality (neutral bulk temperature) $t_{b,n}$ is, therefore,

$$t_{b,n} = 0.2 \times 33.7 + 0.8 \times 36.8$$

$$= 36.18°\mathrm{C}$$

3. *Heat Loss from Respiration* ($C_{res} + E_{res}$)

Heat loss from respiration is by ''dry'' convective heat transfer due to cool air being inhaled, heated to core temperature in the lungs, and heat transferred in exhaled air to the environment C_{res}. In addition, inhaled air is moistened (to saturation) by the lungs. When exhaled, therefore, there is a mass (heat) transfer from the body core to the outside environment E_{res}. ASHRAE [27] gives the following equations for total respiratory heat loss:

$$C_{res} + E_{res} = [0.0014M(34 - t_a) + 0.0173M(5.87 - P_a)]/A_D$$

A_D is calculated as already described. All other values are basic parameters (P_a in kPa).

V. EXAMPLE CALCULATION

The preceding presentation of the human heat balance equation is in summary form yet still rather cumbersome. The usefulness of the equation in practice is the ability to identify and calculate values showing the relative contribution of each of the components. (This becomes convenient when using a digital computer.) A hypothetical example follows. For example, consider a man standing in a hot environment conducting light work in light clothing.

A. Basic Parameters

For the example the following measurements and estimates have been made to define the human thermal environment:

Air temperature: $t_a = 30°C$
Mean radiant temperature: $t_r = 40°C$
Relative humidity: $\varphi = 60\%$
Air velocity: $v = 0.25 \text{ ms}^{-1}$
Metabolic rate (light work): $M = 100 \text{ Wm}^{-2}$
External work: $W = 0 \text{ Wm}^{-2}$
Clothing: $R_{cl} = 0.0930 \text{ m}^2 \cdot °C \cdot W^{-1}$; $R_{e,cl} = 0.015 \text{ m}^2 \cdot \text{kPa} \cdot W^{-1}$

It is assumed here that R_{cl} and $R_{e,cl}$ have been measured, or are available, from tables for example. It may be more convenient to express the insulation value as a basic parameter in Clo. R_{cl} can then be calculated (in $m^2 \cdot °C \cdot W^{-1}$) using 1 Clo = $0.155 \text{ m}^2 \cdot °C \cdot W^{-1}$ (i.e., in the example, clothing insulation is 0.6 Clo = $0.093 \text{ m}^2 \cdot °C \cdot W^{-1}$). $R_{e,cl}$ values are not widely available, and the example assumes that we know this value. A value of $0.015 \text{ m}^2 \cdot \text{kPa} \cdot W^{-1}$ is, however, typical [27] and could be assumed in some cases. Evaporative resistance may also be described by a variety of clothing indices that can be used directly in the heat balance equation or to calculate $R_{e,cl}$ (e.g., i_m or F_{pcl}).

B. Simple Calculations

$$\text{Metabolic heat production} = M - W = 100 - 0 = 100 \text{ Wm}^{-2}$$

$$f_{cl} = 1 + 0.31 \text{ Clo}$$

or

$$= 1 + \frac{0.31 \, R_{cl}}{0.155} = 1.186$$

For $t_{sk} = 35°C$, saturated vapor pressure at skin temperature is

$$P_{s,sk} = \exp(18.956 - 4030.18/(35 + 235))$$
$$= 56.23 \text{ mb}$$
$$= 5.623 \text{ kPa}$$

For $t_a = 30°C$, saturated vapor pressure at air temperature is

$$P_{s,ta} = \exp(18.956 - 4030.18/(30 + 235)) \text{ mb} = 42.43 \text{ mb}$$
$$= 4.243 \text{ kPa}$$
$$P_a = \varphi\, P_{s,ta} = 0.6 \times 4.243 = 2.55 \text{ kPa}$$

C. Assumptions

$$t_{sk} = 35°C$$
$$\epsilon = 0.95$$
$$A_r/A_D = .77$$
$$LR = 16.5 \text{ K/kPa}$$
$$A_D = 1.8 \text{ m}^{-2}$$

D. Heat Transfer Coefficients

$$v > 0.2 \text{ ms}^{-1}$$

therefore,

$$h_c = 8.3v^{0.6}$$
$$= 3.61 \text{ W} \cdot \text{m}^{-2} \cdot \text{K}^{-1}$$
$$h_e = 16.5\ h_c$$
$$= 59.61 \text{ W} \cdot \text{m}^{-2} \cdot \text{kPa}^{-1}$$

Calculate h_r and t_{cl} using iteration

$$h_r = 4 \times 0.95 \times 0.77 \times 5.67 \times 10^{-8} \times [(t_{cl} + 40)/2 + 273.2]^3$$
$$= 16.59 \times 10^{-8}\ ((t_{cl} + 40)/2 + 273.2)^3$$

$$t_{cl} = \frac{(1/R_{cl}) \times t_{sk} + f_{cl} \times (h_c \times t_a + h_r \times t_r)}{(1/R_{cl}) + f_{cl} \times (h_c + h_r)}$$

$$= \frac{10.75 \times 35 + 1.186(108.30 + 40h_r)}{10.75 + 1.186(3.61 + h_r)}$$

Iteration: Start with $t_{cl} = 0.0$, * TCLOLD takes on the value of t_{cl}.

$$h_r = 16.59 \times 10^{-8}((t_{cl} + 40)/2 + 273.2)^3$$

$$t_{cl} = \frac{10.75 \times 35 + 1.186(108.30 + 40h_r)}{10.75 + 1.186(3.61 + h_r)}$$

Criterion: IF |tcl-TCLOLD| > 0.01 THEN RETURN TO *. This is carried out until the criterion is met. For the example, this provides

$$h_r = 4.99 \text{ W} \cdot \text{m}^{-2} \cdot \text{K}^{-1}$$

$$t_{cl} = 35.4°\text{C}$$

Operative temperature:

$$t_o = \frac{h_r t_r + h_c t_a}{h_r + h_c}$$

$$= \frac{4.99 \times 40 + 3.61 \times 30}{4.99 + 3.61}$$

$$= 35.8°\text{C}$$

Combined heat transfer coefficient:

$$h = h_c + h_r$$

$$= 4.99 + 3.61$$

$$= 8.60 \text{ W} \cdot \text{m}^{-2} \cdot \text{K}^{-1}$$

E. Calculation of the Components of the Heat Balance Equation

$$(C + R) = \frac{t_{sk} - t_o}{R_{cl} + 1/(f_{cl} \cdot h)}$$

$$= \frac{(35 - 35.8)}{0.093 + 1/(1.186 \times 8.6)}$$

$$= -4.18 \text{ Wm}^{-2}$$

(i.e., a heat gain).

$$E_{sk} = \frac{w(P_{sk,s} - P_a)}{R_{e,cl} + 1/(f_{cl} h_e)}$$

$$= \frac{w(5.623 - 2.55)}{0.15 + 1/(1.186 \times 59.61)}$$

$$= w \times 105.58 \text{ Wm}^{-2}$$

For $w = 1$, $E_{sk} = E_{max}$, so

$$
\begin{aligned}
E_{max} &= 105.58\ \text{Wm}^{-2} \\
(C_{res} + E_{res}) &= [0.0014M(34 - t_a) + 0.0173\text{M}(5.87 - P_a)]/A_D \\
&= [0.0014 \times 100 \times (34 - 30) + 0.0173 \times 100\,(5.87 \\
&\quad - 2.55)]/1.8 \\
&= 3.50\ \text{Wm}^{-2}
\end{aligned}
$$

The heat balance equation then becomes

$$
\begin{aligned}
(M - W) &= (C + R) + E_{sk} + C_{res} + E_{res} \\
(100 - 0) &= -4.18 + w \times 105.58 + 3.50
\end{aligned}
$$

Thus, a skin wettedness of 0.95 will provide sufficient heat loss at the skin through evaporation. That is, the body will sweat to thermoregulate and achieve heat balance. For maximum evaporation E_{max}, wettedness is 1.0. For the example, this gives a value of $E_{max} = 105.58\ \text{Wm}^{-2}$. The required wettedness for balance is

$$
W_{req} = \frac{E_{req}}{E_{max}}
$$

Therefore, $E_{req} = W_{req} \cdot E_{max} = 0.95 \times 105.58 = 100.67\ \text{Wm}^{-2}$ (i.e., E_{sk} for heat balance).

Using the latent heat of vaporization of water ($22.5 \times 10^{-5}\ \text{Jkg}^{-1}$) and taking account of the efficiency of sweating r (e.g., some sweat drips and latent heat is not lost), ISO 7933 uses

$$
r = 1 - w^2/2
$$

the sweating required (to provide the evaporation required) can then be calculated as

$$
S_{req} = E_{req}/r \qquad (\text{Wm}^{-2})
$$

where an evaporated sweat rate of 0.26 liters (kg) per hour corresponds to a heat loss of 100 Wm^{-2}. For the example, to maintain heat balance the body would therefore have to produce 0.476 liters of sweat per hour; i.e.,

$$
\begin{aligned}
r &= 1 - (0.95)^2/2 \\
&= 0.55 \\
S_{req} &= 100.67/.55 \\
&= 183.04\ \text{Wm}^{-2} \\
&= 0.476\ \text{Lhr}^{-1}
\end{aligned}
$$

This example demonstrates a calculation of the heat balance equation and how it can be used to identify the relative importance of each of the components. Although sweating required in a hot environment is used in the example, requirements for other components can also be calculated. For example, metabolic rate, air velocity, or the insulation of clothing can also be calculated.

The conceptual basis for the body heat balance equation is well established. However, equations for components of the equation are continually updated from the results of research. Of particular importance has been the theory of heat transfer. Much of the development of the heat balance equation for the human body has been adapted from heat (and mass) transfer theory. An understanding of this theory is useful for a thorough understanding of the body heat balance equation. A computer program that allows the calculation of the body heat balance equation given as an example in this chapter, is provided in Parsons [30].

VI. DETERMINATION OF THE THERMAL PROPERTIES OF CLOTHING

Tests are available for determining specific properties of clothing, using heated flat plates or cylinders, for example, to determine I_{cl}, I_{ecl} values (e.g., [31,32]). A comprehensive program to design and evaluate clothing will involve a number of tests and trials. Laboratories specializing in clothing science have developed integrated methods for assessing clothing.

A number of authors have proposed the "clothing triangle" as a method for developing and evaluating clothing (Figure 6). The wide base of the triangle represents the wide range of simple tests that are performed on fabrics using simple heat transfer apparatus (i.e., they are easy to conduct and repeatable, but they are unrealistic as they do not use human subjects). The narrow peak of the triangle represents field evaluation trials using humans wearing the clothing. The methods require relatively large resources and are difficult to control, but they are realistic. Goldman [33] presents five levels of clothing evaluation. Apparatus methods include level 1, the physical analysis of materials, and level 2, the biophysical analysis of clothing ensembles (e.g., using manikins) and predictive modeling. Levels 3, 4, and 5 involve human subjects and involve controlled climatic chamber tests, controlled field trials, and field evaluations, respectively. As the costs of the tests increase with increasing level, it is important to use information from lower levels in the planning of tests at a higher level.

Umbach [31] summarizes the work of the Hohenstein Institute in Germany in a five-level system shown in Figure 6.

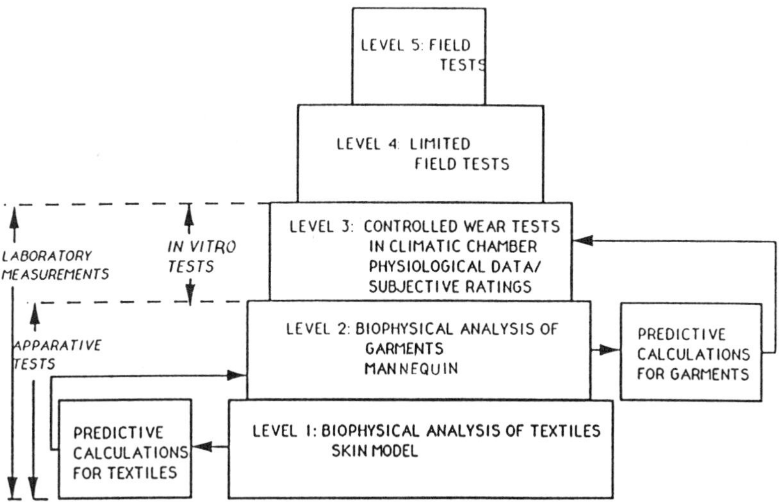

Figure 6 A five-level system for the physiological properties of textiles and garments. (From Ref. 31.)

VII. USER TESTS AND TRIALS

User performance tests and trials of clothing can be relatively expensive to conduct; however, they provide realism. A simple deficiency of any system not involving human subjects would be in the interaction of clothing insulation with human activity. People do not only work, stand, or lie down. They bend, run, and sit and at different rates and angles. These are numerous factors where humans differ from manikins. Using human subjects to evaluate clothing will reduce control therefore, but it is the only way to provide a realistic and comprehensive evaluation.

The degree of control depends upon the use of the human subjects. User performance tests for example, in climatic chambers can use human subjects as manikins to provide measurements under controlled conditions. User trials can involve human subjects wearing the clothing from day to day under normal operating conditions.

A. User Performance Tests

An example of user performance tests to determine the thermal properties of clothing is provided by Parsons [8]. An initial test involves eight male subjects standing stationary in a climatic chamber of air and mean radiant temperature of 5°C, 50%rh, and still air. Two sessions are required: one where the subject is clothed and the other, minimally clothed (shorts and shoes only). The difference, in mean, mean skin temperature (over subjects), after 1 hr of exposure, between the two conditions is given a clothing index value dt_{sk}. Using a model of human thermoregulation [21], an I_{cl} value can be calculated, that is, an I_{cl} value for the model that would give a similar fall in mean skin temperature over time to that of the mean fall for human subjects. This method was extended [34] to include the effects of pumping. Subjects are exposed to a further 1-hr session where they move their arms and legs alternatively and slowly in wide circles. This provides maximum pumping, and the model of human thermoregulation again can be used to calculate clothing insulation. The resultant insulation of the clothing can then be determined from

$$I_{res} = I_{cl} - \frac{A\Delta I_{cl}\,(33 - T_a)\ \mathrm{Clo}}{28} \qquad \text{for} \quad T_a > 5°\mathrm{C}$$

where

I_{cl} = intrinsic clothing insulation
ΔI_{cl} = maximum fall in clothing insulation due to pumping at 5°C
T_a = air temperature
A = activity factor, 0 (for standing subject) to 1 (for maximum pumping)

The method appears to work well and provide realistic values from human subjects. However, it is possible that results are confounded by some (thin) subjects who may shiver at 5°C, and it is important to provide good control over the environment. An advantage of this and similar methods is that the data are realistic. It clearly demonstrates wide individual differences and can quantify these. Further development is required; however, it is possible that such methods may provide preferred alternatives to the use of manikins. Hollies [35] described standardized tests of the assessment of clothing. Human female subjects provide subjective judgments of clothing while in a thermal chamber that provides transient changes in air temperature and humidity. The system has been used for laboratory testing of consumer clothing. User performance tests vary in kind of control and realism; however, none consider clothing under actual use.

B. User Performance Trials

The principle of the user performance trial is to investigate clothing while in actual use, hence providing practical information. This will involve identifying a sample of users of the clothing and observing the properties over a period of time representing realistic conditions. Questionnaire techniques and possibly some physiological measures can be used. For example, in a final evaluation of a clothing ensemble, the clothing will be given to a sample (or all) of users for a period of time (e.g., days, weeks, or months). The users may be observed at frequent intervals, interviewed, or asked to complete questionnaires. If experimenter presence is not possible or desirable, then the subjects may be asked to complete a diary of their experiences. A more convenient method may be to ask the subjects to complete a questionnaire at the end of the trial period. Examples of user trials would include giving a sample of consumers a new type of shirt or sports clothing. Specific user tests and trials may be developed for fire fighter's clothing. When introducing new protective clothing into a steel mill after careful laboratory testing, for example, gradual introduction in carefully controlled trials may be necessary. This should take account of the total role of the clothing in the work system, and good public relations will be important for effective feedback on performance. Similar and extensive trials are also conducted for military use. Tests of new sleeping bags or chemical protective clothing, for example, are performed in both laboratory and field trials.

The measurement techniques used will depend upon the clothing application. An evaluation of boots, for example, may include a measure of the frequency and severity of blisters as well as questions concerning sweating or cold feet.

A comparison of different clothing ensembles is often required, and physiological measures are useful. However, a user trial, although confounded with many factors that cannot be controlled, should be conducted using correct techniques of experimental design. This will maximize useful information obtained. A discussion of field evaluation methods for military clothing is provided by Bechamann [36]. It becomes apparent when considering field trials for the thermal properties of clothing that any evaluation will be context dependent. Isolating only thermal properties will be difficult, and they will interact with many other factors; for example, ventilation properties, bulk, and fit (allowing ease of movement) will all interact. When considering the thermal properties of clothing in user trials, therefore, it will be necessary to consider the wider context.

VIII. THE DESIGN OF FUNCTIONAL CLOTHING

The development of the "ergonomics systems" methodology (e.g., [37]) has led to a change of philosophy in designing and evaluating "components" of a system. Clothing can be seen, therefore, in terms of a dynamic component of an overall man–machine–environment–organization system. Despite the gen-

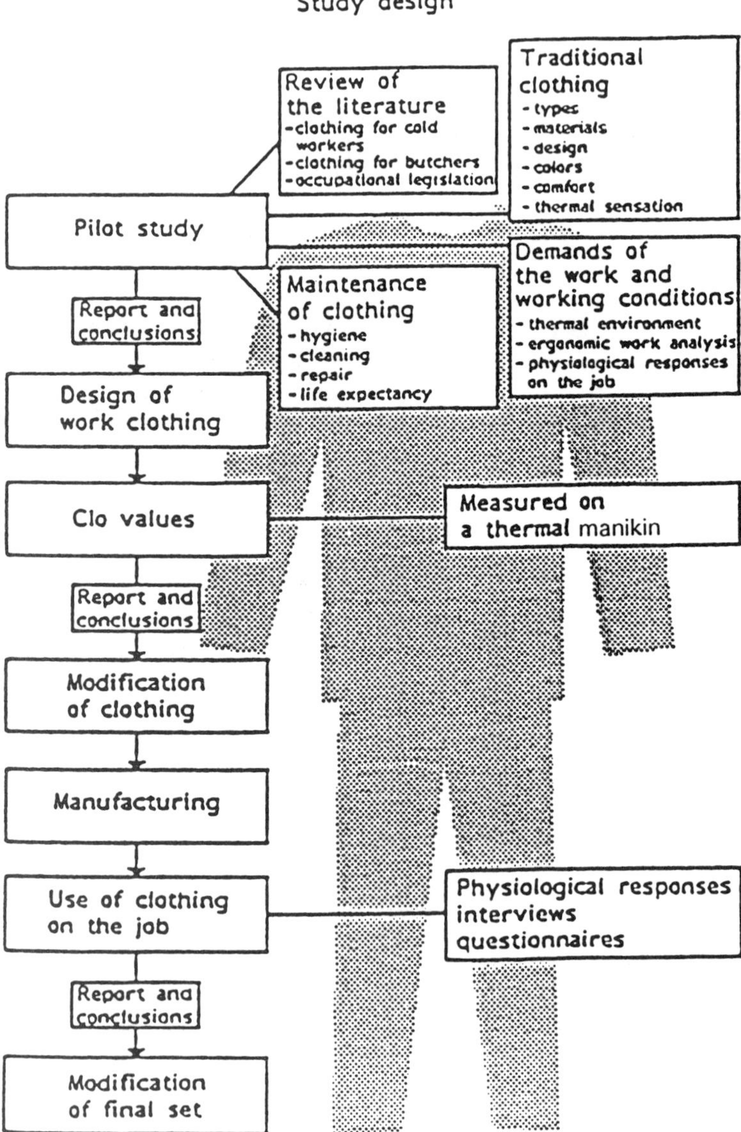

Figure 7 Systems approach to the design, development and implementation of clothing. (From Ref. 40.)

eral acceptance of this philosophy, clothing design and assessment has been slow to use it in practice. For a more complete description of systems ergonomic and practical ergonomics methodology, the reader is referred to Singleton [37] and Wilson and Corlett [38].

Ilmarinen et al. [39] use a simple systems approach to the design, development, and evaluation of meat cutters' clothing. An interesting point is that a team of investigators is required for this type of study involving physiologists, designers, and ergonomists. The tasks and problems of the meat cutter were studied in depth, and clothing requirements were identified. This involved literature reviews, questionnaires and interviews, work analysis, and physiological measurements. Particular attention was paid to the activity of the meat cutters and the thermal and general working conditions in slaughter houses. The overall systems approach to the design, development, and implementation of the clothing is shown in Figure 7.

Even with this relatively comprehensive study, only the simple model of dry thermal insulation I_{cl} is used to specify clothing requirements for thermal environments. More complex thermal models and their component values (I_{cl}, I_{ecl}, etc.) and general methodologies provide further practical information. However there is some way to go before all "ingredients" are linked to provide a comprehensive thermal design and evaluation methodology for clothing. Cost and convenience will always play a role, but the benefits of optimum design of clothing should not be underestimated.

REFERENCES

1. A. C. Burton and O. G. Edholm, *Man in Cold Environment*, Edward Arnold Publishers Ltd., London, 1955.
2. A. P. Gagge, A. C. Burton, and H. C. Bazett, A practical system of units for the description of the heat exchange of man with his environment, *Science 94*:428 (1941).
3. F. T. Pierce and W. H. Rees, The transmission of heat through textile fabrics, Part II, *J. Textile Inst. 37*:181 (1946).
4. E. A. McCullough and B. W. Jones, *A Comprehensive Database for Estimating Clothing Insulation*, IER Tech Report 84-01, Institute for Environmental Research, Kansas State University, 1984.
5. B. W. Olesen, E. Sliwinska, T. L. Madsen, and P. O. Fanger, Effect of body posture and activity on the thermal insulation of clothing: Measurements by a moveable thermal manikin, *ASHRAE Trans. 88* (2):791 (1982).
6. P. M. J. Jones and G. West, The Loughborough Anthropometric Shadow Scanner LASS, 1990.
7. P. O. Fanger, *Thermal Comfort*, Danish Technical Press, Copenhagen, 1970.
8. K. C. Parsons, Protective clothing: heat exchange and physiological objectives, *Ergonomics 31* (7):991–1007 (1988).

9. T. Oohori, L. G. Berglund, and A. P. Gagge, Comparison of current two parameter indices of vapour permeation of clothing—As factors governing thermal equilibrium and human comfort, *ASHRAE Trans. 90* (2A):85 (1984).
10. D. McK. Kerslake, *The Stress of Hot Environment*, Cambridge, 1972.
11. D. P. Wyon, C. Tennestedt, I. Lundgren, and S. Larsson, A new method for the detailed assessment of human heat balance in vehicles—Volvo's Thermal Manikin, VOLTMAN, SAE Technical Paper Series No. 850042, 1985.
12. ISO 9920, *Estimation of the Thermal Characteristics of a Clothing Ensemble*, International Standards Organisation, Geneva, 1992.
13. B. W. Olesen and F. N. Dukes-Dobos, International standards for assessing the effect of clothing on heat tolerance and comfort, *Performance of Protective Clothing*, ASTM STP 989 (Mansdorf, Sager, and Nielsen, eds.), American Technical Publishers Ltd., 1988.
14. C. H. Sprague and D. M. Munsen, A composite ensemble method for estimating thermal insulating values of clothing, *ASHRAE Trans. 80* (1):120 (1974).
15. R. D. Parker and K. C. Parsons, Computer based system for the estimation of clothing insulation and metabolic heat production, *Contemporary Ergonomics 1990* (E. J. Lovesey, ed.), Proceedings of the Ergonomics Society's Annual Conference, Leeds, England, 3–6 April 1990, Taylor and Francis, London, 1990, pp. 473–478.
16. K. C. Parsons, *Human Thermal Environments*, Taylor and Francis, London, 1993.
17. R. A. Haslam and K. C. Parsons, Quantifying the effects of clothing for models of human response to the thermal environment, *Ergonomics 31* (12):1787 (1988).
18. D. A. McIntyre, *Indoor Climate*, Applied Science Publishers, Ltd., London, 1980.
19. T. Oohori, L. G. Berglund, and A. P. Gagge, Comparison of current two parameter indices of vapour permeation of clothing—As factors governing thermal equilibrium and human comfort. *ASHRAE Trans. 90*, part 2A: 85–101 (1984).
20. A. H. Woodcock, Moisture transfer in textile systems, part 1, *Textile Research Journal 32*:628 (1962).
21. Y. Nishi and A. P. Gagge, Moisture permeation of clothing—A factor governing thermal equilibrium and comfort, *ASHRAE Trans. 76* (1):137 (1970).
22. Y. Nishi, and A. P. Gagge, Moisture permeation of clothing—A factor governing thermal equilibrium and comfort, *ASHRAE Trans. 76*, part 1:137–145 (1970).
23. W. A. Lotens and E. J. G. v. d. Linde, Insufficiency of current clothing description, *Conf. Med. Biophys. Aspects of Protective Clothing*, 1983.
24. E. A. McCullough, B. W. Jones, and T. Tamura, A data base for determining the evaporative resistance of clothing, *ASHRAE Trans. 95* (1989).
25. W. A., Lotens, A clothing model. Proceedings of the International Conference on Environmental Ergonomics, Austin, Texas, 1989.

26. D. DuBois and E. F. DuBois, A formula to estimate surface area, if height and weight are known, *Archives of Internal Medicine 17*:863 (1916).
27. ASHRAE, Physiological principles, comfort and health, *Fundamentals Handbook*, American Society of Heating, Refrigeration and Air-Conditioning Engineers, Atlanta, 1989.
28. D. Mitchell, Convective heat loss from man and other animals, *Heat Loss from Animals and Man* (J. L. Monteith and L. E. Mount, eds.), Butterworths, London, 1974.
29. P. O. Fanger, Calculation of thermal comfort: Introduction of a basic comfort equation, *ASHRAE Trans. 73,* part 2 (1967).
30. K. C. Parsons, The Thermal Audit: Report of Human Modelling Group, Loughborough University, 1993.
31. K. H. Umbach, Physiological tests and evaluation models for the optimization of the performance of protective clothing, *Environmental Ergonomics* (I. B. Mekjavic, E. W. Banister, and J. B. Morrison, eds.), Taylor and Francis, London, 1988, pp. 139–161.
32. E. A. McCullough, B. W. Jones, and T. Tamura, A data base for determining the evaporative resistance of clothing. *ASHRAE Trans. 95* (1989).
33. R. F. Goldman, Standards for human exposure to heat, *Environmental Ergonomics* (I. B. Mekjavic, E. W. Banister, and J. B. Morrison, eds.), Taylor and Francis, London, 1988.
34. K. C. Parsons, The thermal audit, *Contemporary Ergonomics,* (E. J. Lovesey, Ed.), Taylor and Francis, London, 1992.
35. N. R. S. Hollies, A. G. Custer, C. J. Morin, and M. E. Howard, A human perception analysis approach to clothing comfort, *Textile Research Journal* 49:557–564 (October 1979).
36. F. W. Behamann, Field evaluation methods, *Handbook of clothing. Research Group 7: Biomedical Research Aspects of Military Protective Clothing*, NATO, Brussels, 1988.
37. W. T. Singleton, *Man–Machine Systems,* Penguin, London, 1974.
38. J. R. Wilson and E. N. Corlett, *Evaluation of Human Work—A Practical Ergonomics Methodology*, Taylor and Francis, London, 1990.
39. R. Ilmarinen, E. Tammela, and E. Korhonen, Design of functional work clothing for meat cutters, *Applied Ergonomics 21* (1):2 (1990).

7

Protective Clothing: Use of Flame-Retardant Textile Finishes

THERESA A. PERENICH The University of Georgia, Athens, Georgia

I. INTRODUCTION

Flammability of textile materials has been recognized as a major cause of burn injuries throughout the world. In 1953 and 1967, the concern over flammability of apparel items worn in the home prompted legislation (Flammable Fabrics Act and Amended Flammable Fabrics Act) to introduce mandatory flammability standards that would protect the public against unreasonable risk. In 1972, the Consumer Product Safety Act was enacted and the Consumer Product Safety Commission (CPSC) was established with broad jurisdiction over consumer safety. However, the concern over flammability of garments is not limited to the consumer alone. Federal, state, and local governments; private and public industries; and businesses are involved in providing their personnel with protection and protective apparel for fire situations. Intensive effort has been expended in the design, development, and fabrication of suitable and reasonably priced flame-retardant fabrics for apparel for the military, refinery, and fire brigade sectors as well as for workers in paints and resin plants, utilities, steel mills, aluminum mills, and others.

In fire situations, fabrics are subjected to extremely heavy loads. Fabric performance in these situations is related to comfort, time, heat, durability, and other characteristics specific to the occurrence. Stricter regulations by the National Fire Protection Association (NFPA) and the National Institute for Occupational Safety and Health (NIOSH) have targeted the need for specific performance characteristics for flame-retardant fabrics. These characteristics include maximum and constant protection against heat and fire, high heat insulation, high strength, optimum dimensional stability, high abrasion resistance, excellent hand and wearing qualities, and optimum dyeability [1].

In examining methods to impart flame resistance on fabrics, textiles are frequently classified into four groups: highly flammable, burn with great difficulty, burn easily, and flame resistant. To impart flame resistance, fabrics are produced by either using inherently flame-resistant fibers or using fiber variants that are flame resistant by the addition of flame retardants in the spinning solution or by the application of flame-retardant finishes to the fabrics. This chapter will focus on flame-retardant finishes and their application and uses.

II. TEXTILE FLAMMABILITY REGULATIONS AND STANDARDS

A number of federal agencies involved in fire protection are listed in Table 1 [2].

Regulations and standards have been developed and are required for a variety of products, including building materials and plastics in addition to textiles. Those agencies dealing with textile flammability standards and regulations include the Department of Commerce, Consumer Product Safety

Table 1 Selected Federal Agencies Involved in Fire Protection

Agency/Department	Organization
Consumer Product Safety Commission (CPSC)	
Department of Agriculture	U.S. Forest Service (USFS)
Department of Commerce	Economic Development Administration (EDA)
	Maritime Administration (MARAD)
	National Institute of Standards and Technology (NIST)
	National Fire Prevention and Control Administration (NFPCA)
Department of Defense (DOD)	Civil Defense Preparedness Agency (CDPA)
	Military Department (Air Force, Army, Marine Corps, Navy)

Table 1 (Continued)

Agency/Department	Organization
Department of Health, and Human Services (HHS)	National Institutes of Health (NIH) National Institute of Occupational Safety and Health (NIOSH) Public Health Service (PHS) Social Security Administration (SSA)
Department of Housing and Urban Development (HUD)	Federal Disaster Assistance Administration Federal Housing Administration (FHA)
Department of the Interior	Bureau of Land Management (BLM) Bureau of Mines (BUMINES) Mining Enforcement and Safety Administration (MESA) National Park Service (NPS)
Department of Labor (DOL)	Bureau of Labor Statistics (BLS) Occupational Safety and Health Administration (OSHA) Wage and Manpower Division
Department of Transportation (DOT)	Federal Aviation Administration (FAA) Federal Railroad Administration (FRA) Materials Transportation Bureau (MTB) National Highway Traffic Safety Administration (NHTSA)
Environmental Protection Agency (EPA)	Federal Trade Commission (FTC)
General Services Administration (GSA)	Federal Supply Service
National Academy of Engineering (NAE)	
National Academy of Sciences (NAS)	
National Aeronautics and Space Administration (NASA)	
National Research Council	
National Science Foundation (NSF)	
National Transportation Safety Board (NTSB)	
Nuclear Regulatory Commission (NRC)	
Small Business Administration (SBA)	

Source: Adapted from Ref. 2.

Commission, National Fire Protection Agency, Federal Aviation Authority, and the American Society for Testing and Materials.

III. TEST METHODS

A. Development and Generation of Test Methods: Organizations

Specific organizations are involved in the development and generation of test methods. One of the oldest and most recognized organizations generating test methods is the American Society for Testing and Materials (ASTM), which was founded in 1898. It is a scientific and technical organization formed to develop standards on characteristics and performance of materials, products, systems, and services and to promote related knowledge [3]. The ASTM is the world's largest source of voluntary consensus standards and operates through more than 140 main technical committees with 1,944 subcommittees. There are over 28,000 active members in ASTM, and an ASTM standard represents a common viewpoint of producers, consumers, users, and general interest groups who are concerned with its provisions. The ASTM standards are designed to assist industry, government agencies, and the general public. Standards generated by ASTM are reviewed periodically, and the use of the standards is voluntary. The ASTM Committee E-5 deals with fire standards related to fire performance of materials, products, and systems within their relevant environment. Specifications on textile flammability are part of ASTM Committee D-13 on Textiles. Table 2 lists the test methods related to flammability of textiles and other materials.

Before a product or system may be considered eligible for Listing, Classification, Recognition, or Certificate Service by Underwriters Laboratories, it must meet technical requirements developed and generated internationally by the organization. Underwriters Laboratories is the governing body for its standards, and its engineers revise and generate new standards based on input from safety experts, standardization experts, vendors, manufacturers, insurance personnel, and other interested parties. The Tunnel Test for Building Materials is a standard that is cross referenced with an ASTM specification (ASTM E 84) [2].

The National Fire Protection Association (NFPA) was started in 1896 with a focus on promoting the science, improving the methods of fire protection, and obtaining and circulating information on fire protection. The NFPA is a private, voluntary organization with over 2,400 members with activities in information exchange, research, technical advisory services, public education, technical standards generation, and services to public protection agen-

Table 2 Test Methods Related to Flammability of Textiles and Other Materials

Test method	Test title	Applicable material	Properties measured	Specimen angle
Textiles				
ASTM E648/NFPA 253/FTS 372 (Flooring Radiant Panel)	Critical Radiant Flux of Floor covering systems	Floor covering systems	Critical radiant flux at flame out	Horizontal
ASTM D 2859 (Pill Test)	Surface Flammability of Carpets and Rugs	Carpets and rugs	Area of flame spread greatest diameter	Horizontal
DOC FF-1-70 (Pill Test) CPSC 16 CFR 1630.4	Surface Flammability of Carpets and Rugs	Carpets and rugs	Area of flame spread greatest diameter	Horizontal
FED. STD. DDD-C-95 (Pill Test)	Surface Flammability of Carpets and Rugs	Carpets and rugs	Area of flame spread greatest diameter	Horizontal
FED. STD. 191 Method 5900	Flame Resistance of Cloth; Horizontal	Fabrics	Area of flame spread greatest diameter	Horizontal
FED. STD. 191 Method 5903	Flame Resistance of Cloth; Vertical	Fabrics	Char length, after flame, after glow	Vertical
FED. STD. 191 Method 5906	Burning Rate of Cloth; Horizontal	Fabrics	Rate of burning	Horizontal
FMVSS 302	Flammability of Interior Materials—Cars, Trucks, Multipurpose Passenger Vehicles, Buses	Textile fabrics	Rate of burning	Horizontal
NFPA 701 (Large Scale)	Fire Tests for Flame Resistance Textiles and Films	Fabrics and films	Char length, after flame, after glow	Vertical

Table 2 (Continued)

Test method	Test title	Applicable material	Properties measured	Specimen angle
NFPA 701 (Small Scale)	Fire Tests for Flame Resistance Textiles and Films	Fabrics and films	Char length, after flame, after glow	Vertical
Federal Aviation Authority Para. 25.853	F.A.A. Regulations for Compartment Interiors	Compartment materials	Time of flame spread	Horizontal
Federal Aviation Authority Para. 25.853	F.A.A. Regulations for Compartment Interiors	Compartment Materials	After flame, burn length	Vertical
Plastics				
ASTM D 568	Flammability of Plastics 0.050″ and Under	Plastic sheets and films	Burn length, burn rate	Vertical
ASTM D 635	Flammability of Plastics Over 0.050″ Thickness	Rigid plastics	Burn rate, extent of burning	45°
ASTM D 1929 Proc. B	Ignition Properties of Plastics	Plastic material	Flash and self-ignition temperatures	Horizontal
ASTM D 2863	Oxygen Index Flammability Test	Plastics and textiles	Oxygen index value	Vertical
CPSC 16 CFR 1500.44	Hazardous Substance Labeling Act (Toys)	Rigid and pliable bonds	Rate of burning	Horizontal
Miscellaneous Building Materials				
ASTM E 84/NFPA 255/UL 723 (Tunnel Test)	Surface Burning of Building Materials	Building materials	Flame spread, smoke developed	Horizontal

ASTM E 119	Fire Tests of Building Construction and Materials	Wall assemblies	Fire penetration, structural integrity	Vertical
ASTM E 136	Noncombustibility of Elementary Material	Building materials	Noncombustibility of Materials	Vertical
ASTM E 152	Fire Test of Door Assemblies	Door and window assemblies	Fire penetration, structural integrity	Vertical
ASTM E 162 (Radiant Panel)	Surface Flammability of Materials Using a Radiant Heat Energy Source	Building materials	Flame stread index	60°
ASTM E 603	Standard Guide to Room Fire Experiments	Building materials products or systems	Ignition time, flash over, flame growth, rate of smoke development	Actual orientation
ASTM E 662/NFPA 258 (Aminco Smoke Den. Chamber)	Specific Optical Density Smoke	All types of insulation	Specific optical density	Vertical
CPSC 16 CFR PART 1209.6 (Insulation Radiant Panel)	Safety Standard for Cellulose Insulation	Insulation	Critical radiant flux at flame out	Horizontal
ANSI 634/IEEE 634	Cable Penetration Fire Stop Qualification	Cable penetration fire stop assemblies	Fire penetration structural integrity	Vertical
Corner Test	Full or Small Scale Corner Geometrics	Building materials and assemblies	Fire propagation, fire involvement	Vertical
Toxicity	NBS Toxicity Protocol	Building materials products or systems	Toxicity of smoke	N/A

cies [2]. Through its information exchange activity, NFPA publishes several journals, handbooks, texts, audiovisual materials including *Fire Journal*, *Fire Technology*, *Fire News*, and the *NFPA Fire Protection Handbook*. Technical standards are generated by over 150 committees, which are composed of members with an interest in the standards. Two test methods for textile flammability under the designation NFPA 701 (large and small scale) have been generated by NFPA to measure char length, after flame, and after glow on fabrics and films (Table 2). Fire fighters' turnout clothing material and fire fighters' station/work uniforms are specified in NFPA 1971 and NFPA 1975, respectively. Specifications for clothing for personnel working with molten substances are described in ASTM F 1002 [3].

In addition to ASTM, NFPA, and Underwriters Laboratories, the American National Standards Institute (ANSI), insurance underwriters, and federal agencies also develop test specifications. Two agencies within the Department of Commerce that are concerned with fire safety are the National Fire Prevention and Control Administration and the National Institute of Standards and Technology. The Occupational Safety and Health Administration under the Department of Labor provides standards for the safety and health of workers in their places of employment.

B. Test Methods Related to Flammability

Table 2 presents test methods related to fire and flammability testing for textiles, plastics, and some building materials. The most widely used test methods developed for determining the flammability of construction, building materials, and plastics are those developed by ASTM. There are over 15 fire test methods encompassing combustible properties of wood, surface burning characteristics of building materials, and fire tests for roof coverings, to name a few. These specific test methods are under the auspices of ASTM Committee E-5 and are described in Section 4 of the *Annual Book of ASTM Standards*. Although these particular test methods are beyond the scope of this chapter, it is important to note that flammability of many different products is a concern and that test methods and standards have been generated to protect property and persons from fire related accidents.

The concern over deaths and injuries due to fire situations involving textile products culminated in the promulgation of the Flammable Fabrics Act of 1953. This legislation was intended to protect consumers from hazardous apparel. In December 1967, the Act was amended to include all personal and household fabrics and all textile and textile type products. In 1970, flammability specifications were issued covering carpets and rugs. In 1971, a children's sleepwear flammability standard for sizes 0 to 6X was promulgated. In

1972, flammability specifications for mattresses were issued, and in 1974, the Federal Register published the Standard for the Flammability of Children's Sleepwear, sizes 7 through 14 (FF 5-74), which was issued under the Flammable Fabrics Act.

Flammability tests for fabrics can be grouped into three general categories depending on the manner in which the sample is tested. As a general rule, the most severe test for most fabrics is one in which the sample is tested in a vertical position. For a variety of fabrics, horizontal testing is less severe, and inclined sample testing (45° angle) is the least severe of the tests. In most test methods, the source of ignition is a flammable gas such as methane, and place of ignition is dependent on the positioning of the fabric sample. The properties that are measured during flammability testing are ease of ignition, burn rate, flame spread, ease of extinguishment, smoke density, rate of heat release, and thermal resistance.

1. *Vertical Testing*

In this method, conditioned fabric samples are suspended vertically in holders in a steel cabinet of specified dimensions. The test specimen is exposed to a gas flame along its bottom edge for 3 sec and under controlled conditions. Char length and residual flame time are measured. Specimens fail the test if the char length is over 7 in. Vertical testing is used for DOC FF3-71 and DOC FF5-74, which require that any product of wearing apparel up to and including sizes 0 to 6X and 7 to 14, such as nightgowns, pajamas, or other items intended to be worn for sleeping must pass the test. This requirement excludes diapers and underwear, and the items must meet the standard after 50 launderings and dryings. Federal Test Method Standard 191a, Methods 5902, 5903 and 5904, and NFPA 701 also use vertical testing to determine the flame resistance of fabrics used for fire fighters' clothing and clothing of other workers who are exposed to open flame, flammable atmospheres, etc.

2. *Diagonal (45° and 30°) Testing*

A diagonal flame test apparatus is used to determine the flame resistance of a textile according to Federal Test Method Standard No. 191, Method 5908. In this method, a fabric specimen is mounted at a 45° angle and is exposed to an open flame for a specified period of time. The rate of burning and ease of ignition of apparel fabrics are measured.

3. *Horizontal Testing*

The most common horizontal tests for textiles and clothing are described by Federal Specification 191, Methods 5900 and 5906. For horizontal testing, the conditioned fabric specimen is mounted in a horizontal holder and exposed to an open flame for a specified period of time in a horizontal flame test apparatus. The burning rate and char hole diameter of the sample are measured after a specified period of time. The horizontal method of testing fabrics is applicable to untreated, flame-proofed, and fire-resistant fabrics. The Department of Transportation uses the horizontal flame test to determine the flammability of automotive interiors.

C. Other Flammability Tests

In addition to those tests briefly described, other procedures have been developed to assess the flame resistance of textiles.

1. Methenamine Test

The Methenamine Pill Test was developed to test the flame resistance of carpets and rugs (DOC FF-1-70 and DOC FF-2-70). A methenamine tablet is ignited on a test sample under controlled conditions, and the size of the burn hole is measured.

2. Mushroom Apparel Flammability Test

The Mushroom Apparel Flammability Test involves igniting a cylinder of fabric around a core containing heat sensors. The test measures the rate of heat transfer from the burning material to the heat sensors.

3. Radiant Panel Test

Flammability of carpets and rugs also can be tested using the Radiant Panel Test. In this test, the specimen is mounted on the floor of the test chamber and exposed to intense radiant heat from above. This test measures the rate of flame spread in carpets and rugs.

4. Steiner Tunnel Test

Another test for floor coverings is the Tunnel Test (also called the Steiner Tunnel Test) in which a sample is placed on the ceiling of a tunnel of specific dimensions and ignited under controlled conditions. The test determines the extent to which the floor covering will burn and is required for some carpets that are to be used in nursing home situations.

5. Blanket, Mattress, and Upholstery Tests

The ASTM Method D 4151-82 is a standard test method that is designed to identify blanket fabrics that ignite easily and propagate flame across the surface. Paper monitors are used in this test with the specimen to be tested.

Burning, charring, or discoloration of the paper monitor after ignition classifies the material as sufficiently flammable so that it is unsuitable for blanket manufacture [3].

The cigarette test is used to measure the flammability of mattresses and upholstery fabrics. The Upholstered Furniture Action Council (UFAC) has several fabric classification methods to establish the performance level of upholstery cover fabrics that include flammability testing. The UFAC was formed in 1974 to focus on the problem of a mandatory federal standard for furniture flammability. The UFAC designed a voluntary action program to make upholstered furniture as safe from cigarette ignition as possible. By 1984, 90% of the U.S. furniture market pledged compliance with UFAC.

IV. TYPES OF FLAME-RETARDANT FINISHES

A. History and Development

The potential hazards of flammable fabrics and textiles have received much attention in the past few years. A number of deaths resulting from fires in Great Britain in 1987 has produced legislation on furniture and furnishings and development and use of flame-retardant finishes throughout Europe. It has been estimated that European sales of flame-retardant and soil-release finishes will have doubled in volume from 1986 to 1996 [4]. Currently, in the United States, 30 U.S. suppliers of flame-retardant chemicals for textile uses are listed in the flame retardants buyers' guide of *American Dyestuff Reporter* [5]. There are over 140 flame-retardant products supplied by these companies in liquid, powder, solid, and emulsion forms for use on all types of fibers. The chemicals may be classified as durable, semidurable, or nondurable.

The major markets for fabrics treated with flame-retardant finishes include institutional apparel and fire service areas and iron, steel, aluminum, chemical, oil, petrochemical, and utilities industries [6]. As mentioned earlier, textiles are classified into four groups when consideration is given to impart flame resistance on fabrics. Table 3 presents the four classifications with the burning characteristics of the fibers.

Although fiber is an important factor in predicting the behavior of fabrics in a fire situation, other factors that need to be considered include fabric weight and structure, garment design, finishes applied, and total garment assembly. Other considerations inherent in the use of protective garments in special situations are comfort, durability, launderability (when applicable), maintainability, appearance, and mobility. These aspects form an important critical element in the acceptability of protective apparel by the potential wearer of the items.

Table 3 Burning Behavior of Textile Fibers

Burning characteristics	Fiber	Fiber type
Fuses and shrinks away from flame, melts slowly with burning, self-extinguishing after removal of flame, hard gray or tan ash bead, smells like celery when burning	Nylon	Synthetic polyamide
Fuses and shrinks away from flame, melts when burned, self-extinguishing after flame removal, hard tan or off-white ash bead, has blue or yellow flame	Olefin	Synthetic long chain polymer (at least 85%, by weight, ethylene, propylene)
Fuses and shrinks away from flame, burns slowly with melting, black smoke, usually self-extinguishing after removal from flame, hard black ash bead, sweetish odor	Polyester	Synthetic, long-chain polymer (at least 85%, by weight, ester of substituted aromatic carboxylic acid)
Curls away from flame, difficult to ignite, burns weakly when removed from flame, often self-extinguishes, crushable black ash, burning hair odor	Silk	Natural continuous-filament fiber produced by silkworm
Curls away from flame, difficult to ignite, burns weakly when removed from flame, often self-extinguishing, crushable black ash, acrid burning hair odor	Wool	Natural animal hair–fur protein
Fuses away from flame, burns very slowly with melting, normally self-extinguishing, white smoke, brittle hard black ash bead	Modacrylic	Synthetic, modified acrylic, long-chair polymer of <85% but at least 35%, by weight, acrylonitrile units
Fuses but does not shrink away from flame, burns with melting, continues	Spandex	Synthetic, long-chain polymer with at least 85% segmented polyurethane

Table 3 (Continued)

Burning characteristics	Fiber	Fiber type
to burn with melting after removal from flame, soft black ash		
Ignites and burns easily, does not shrink away from	Cotton	Natural cellulosic; from cotton plant
flame, does not fuse when burned, continues to burn	Flax	Natural cellulosic; from flax plant
after removal from flame, afterglow, gray, feathery, smooth-edged ash, odor of burning paper when ignited	Rayon	Synthetic cellulosic; from wood pulp or cotton linters

Source: Adapted from Ref. 7.

B. Pyrolysis of Textiles

Almost all textile materials will burn when exposed to an ignition source; however, the manner in which the textiles burn is different and depends on several parameters. Terms used to describe the burning behavior of textiles center around the ability of a fabric to resist ignition, to burn slowly, or to self-extinguish after the source of ignition has been removed. The ASTM has provided definitions of several terms used in flammability testing as part of the test methods specified for textiles used for apparel fabrics. The following definitions are provided by ASTM D-123 [3].

Flame resistance: The property of a material whereby flaming combustion is prevented, terminated, or inhibited following application of a flaming or nonflaming source of ignition, with or without subsequent removal of the ignition source.

Flame retardant: A chemical used to impart flame resistance.

Flammable textile: Any combustible textile that burns with a flame.

Flammability: Those characteristics of a material that pertain to its relative ease of ignition and relative ability to sustain combustion.

In addition to those definitions, descriptions of the manner in which textile materials burn are useful.

Flammable textile: One that ignites after exposure to a heat source and continues to burn under normal conditions until it is consumed [8].

Flame-resistant textile: One that will retain its shape and will not ignite after exposure to temperatures up to 600°C [8].

Flame-retardant textile: One that may or may not ignite after exposure to a heat source. If it ignites, it will burn or smolder for only a short time after the ignition source is removed and will self-extinguish [8].

Pyrolysis of textile materials is cyclic, starting with decomposition that occurs when a fabric is exposed to a heat source. As the polymer decomposes, combustibles that ignite in the presence of oxygen to produce flaming combustion are formed. At this point, other volatile combustible materials are formed producing heat. Because the process is cyclic, it may continue until the entire textile material is decomposed producing a char or an ash [8]. Based on the process described previously, flame-retardant chemicals may be characterized as those that are condensed phase active and those that are vapor phase active. Flame-retardant chemicals that are condensed phase active are effective in the substrate and reduce the amount of combustible gases formed. Vapor phase active chemicals are effective in the gaseous combustion stage and reduce the amount of heat generated. The oxidation reactions may be slowed through free radical termination or radical recombination processes [9]. There are two ways to achieve flame resistance of fabrics within this context. One method is to alter the fuel-producing reactions so that no fuel is produced from the polymeric degradation to further the burning. Another method is by interfering with the fuel-consuming process so the heat emitted from the flame is reduced sufficiently where the temperature is too low to maintain fuel-generating reactions [9]. A modified explanation of the two methods is that flame-retardant chemicals may function either in the solid phase where the chemicals lower the decomposition temperature of the fiber and change the nature and amount of volatile combustibles formed or by inhibiting combustion in the vapor phase [8].

Theories that have been proposed and that provide the basis for imparting flame-retardant treatments on textiles include the chemical theory that uses the fact that certain chemicals alter the decomposition of fibers whereby smaller amounts of tars and flammable gases are formed, and the proportion of nonvolatile, carbonaceous materials are increased; the thermal theory in which heat that is supplied may be absorbed by exothermic processes while the flame-retardant chemical is being changed; the gas theory whereby the flame-retardant-treated textile releases an inert gas that interferes with the propagation of the flame; and the coating theory whereby fibers are coated with impermeable coatings that confer flame resistance [2].

Textile flammability is a complex problem because textile materials often are composed of several different polymer types. That substrate may influence the combustion process and configuration, physical environment, and composite characteristic of a textile system and also must be considered. To impart flame-resistant properties to textiles effectively, the decomposition process must be interrupted by the flame-retardant chemical.

Additional factors that contribute to the complexity of textile flammability are additives and other finishes that are applied to the fabric to provide soil release, durable press, and other properties and the high surface-to-volume ratio that characterizes textiles [10].

V. CLASSIFICATION OF FLAME-RETARDANT CHEMICALS

A. Durable, Semidurable, and Nondurable

Durable flame-retardant finishes react with or are physically contained on the surface of the fabric or are incorporated into the fiber. These finishes should last the entire useful life of the fabric and not be removed in cleaning or laundering processes. The application of durable flame-retardant finishes is done by a pad-dry-cure process in which the finish formulation contains the flame-retardant chemical as well as other finishing chemicals such as softeners or crosslinking agents. The finish becomes permanently affixed to the fabric by reacting with the fibers, or it may react with itself to form a polymer within and around the interstices of the fabric and thus be affixed to the fabric without reacting with it. The amount of finish needed to impart flame-retardant properties to a fabric depends on the fiber content, the particular flame retardant being used, the structure of the fabric, and the weight of the fabric. Properties such as strength and abrasion resistance may be lower in fabrics treated with flame-retardant finishes because the amount of add-on of finish is usually between 10 and 30%. The aesthetic qualities of the fabric also may be diminished due to the high add-on needed to impart durable flame-retardant finishes to fabrics.

Semidurable flame-retardant finishes will withstand a limited number of launderings and need to be applied using high curing temperatures to provide the fabric with sufficient flame-retardant capability. Generally, semidurable finishes are applied by padding the chemicals onto the fabric in a two-bath process. In the first bath, the fabric is impregnated with one flame-retardant chemical, and the excess chemical is removed by passing the fabric through the padder. Another flame-retardant chemical is applied in the second bath, which proceeds in the same manner as the previous bath. To obtain satisfactory flame-retardant properties on fabrics using semidurable finishes, a 25 to 30% add-on of chemical is required.

Those flame-retardant finishes that are easily removed in laundering and cleaning are classified as nondurable, and the fabric must be retreated to restore its flame-retardant properties.

B. Flame-Retardant Chemicals

The methods and chemicals used to impart flame-resistant properties to textiles depend on factors such as fiber content of the fabric and the type of treatment desired (i.e., durable, semidurable, nondurable) as discussed previously. These factors determine the effectiveness of the treatment for its end-use performance and are important considerations for the producer and manufacturer of flame-resistant garments and textiles. Flame-retardant chemicals can be grouped into categories specified by their chemical nature. The groupings according to chemical nature are provided in Table 4.

The most widely used flame-retardant chemicals for cellulosic fabrics are those whose active chemical is phosphonium salt: phosphonium salt precondensates, ammonium polyphate, organic borate phosphate, ammonium sulfamate, or organic phosphate. Polyester fibers are treated with phosphonate, high molecular weight polymers, organic phosphate, antimony, organic halogen compounds, bromine–chlorine–phosphorous, emulsified chlorinated compounds, cyclic phosphorous compounds, or organic bromo/phosphate chemicals. Flame-retardant chemicals for blended fabrics include the organic halogens, inorganic salts, organic phosphorous compounds, antimony–

Table 4 Flame-Retardant Chemical Groupings

Fiber	Flame–retardant chemicals used
Cotton	Ammonium polyphate
Flax	Ammonium sulfamate ($NH_4SO_3NH_2$)
Rayon	Organic borate phosphase
	Organic phosphate
	Phosphonium salt
	Phosphonium salt precondensate
Polyester	Antimony
	Chlorinated compounds
	Organic halogen
	Organic phosphate
	Phosphonate
Polyester–cotton	Antimony–bromine compounds
blends	Antimony pentoxide (O_5Sb_2)
	Inorganic salts
	Organic bromo/phosphate
	Organic halogens
	Organic phosphorus compounds

Source: Adapted from Ref. 11.

bromine compounds, boron–phosphate complex, inorganic or organic bromo/phosphate, and antimony pentoxide.

C. Flame-Retardant Finishing of Cotton

Historically, topical finishes of low melting salts such as borax have been used to provide flame resistance on cotton fabrics. However, these finishes are nondurable to laundering, and fabric hand is usually diminished. Flame-retardant finishes for cellulosics have been based on phosphonium compounds, which form crosslinked polymers on fiber surfaces. However, as with topical finishes using salts, the flame resistance is temporary and nondurable to laundering with the additional disadvantage of impairing the hand of the fabric.

Compounds containing phosphorous in combination with nitrogen or a halogen have been widely used as durable flame-retardant chemicals for cotton. Hydroxymethyl phosphines, phosphine oxides, and phosphonium hydroxides have been applied to cotton fabrics since the early 1960s to impart flame-retardant properties. Tris(hydroxymethyl)phosphine (THP), tetrakis(hydroxymethyl)phosphonium hydroxide (THPOH and THPOH-NH_2), tris(hydroxymethyl)phosphine oxide (THPO), and tetrakis(hydroxymethyl)phosphonium chloride (THPC-amide) are some of the most commonly used flame-retardant chemicals used to treat cotton fabrics. In 1970, a flame-retardant treatment based on cyanamide-phosphoric acid that was based on the reaction product of tetrakis(hydroxymethyl)phosphonium chloride (THPC) was developed. Because flammability of textile materials was highlighted during the 1970s, many flame-retardant chemicals were developed and tested for flame resistance and other properties that would make the fabrics useful for consumer products. Some of the most commercially important finishes are:

THP	tris(hydroxymethyl)phosphine $(CH_2OH)_3P$
THPOH	tetrakis(hydroxymethyl)phosphonium hydroxide $(CH_2OH)_4POH$
THPOH-NH_3	tetrakis(hydroxymethyl)phosphonium hydroxide-ammonia $(CH_2OH)_2$-PCH_2NH_2
THPO	tris(hydroxymethyl)phosphine oxide $(CH_2OH)_3O$
THPC-amide	tetrakis(hydroxymethyl)phosphonium chloride $(CH_2OH)_4PC1$

A summary of THP derivatives and curing conditions is presented in Table 5.

Two important considerations in the application of THP finishes that determine the final flame-retardant property and durability are the pH of the

Table 5 Tetrakis(hydroxymethyl)phosphonium Derivatives on Cotton

	THP derivatives	Curing method
(1)	**THPC**	
	THPC-cyanamide	Catalyst–heat 150–160°C
	THPC-dimethylolcyanoguanidine	
	THPC-thiourea	
	THPC-trimethylol melamine (TMM)	
	THPC-urea	
	THPC-tris(1-aziridindyl)phosphine oxide or sulfide (APO or APS)	Catalyst–heat 160°C
	THPC-urea-ammonia	NH_3
(2)	**THPOH** (THPC + NaOH @ pH 7.0–7.5)	
	THPOH-TMM-urea	Heat 150°C or NH_3
	THPOH-ammonia	NH_3
(3)	**THPS**	
	THPS-urea	Catalyst–heat
	THPS-ammonia	NH_3
(4)	**THP carboxylates**	
	THP-acetate (+ phosphate)	NH_3
	THP-oxalate	NH_3

Source: Adapted from Ref. 12.

precondensate and oxidation state of phosphorous in the final polymer. When THPC is heated with cotton, it slowly condenses, and if the cellulose is substituted with amino-ended substituents, substrate affinity may be enhanced [12].

Researchers at the USDA Southern Regional Research Center used sodium phosphate catalysts with a THPC–urea system, and both fabric strength and finish durability were obtained. The Proban process patented and licensed by Albright and Wilson Ltd. is based on the application THPC–urea followed by an ammonia cure. The finish is available as a precondensate of THPC and urea and is padded, dried, and then cured with ammonia, which gives a final ratio of P:N = 1:2 [12]. THPC is usually padded with an amine and/or alkaline medium that partly neutralizes the phosphonium salt acidity. However, THPC-amide finishes are susceptible to hydrolysis, and free formaldehyde may be present. The need to eliminate chlorides from the THPC-amide treatments led to the development of THPOH treatments. Complete neutralization of THPC with NaOH yields THPOH. THPC or THPS (sulfate) reacts with NaOH, and the product of the reactions is an equilibrium mixture or THPOH and THP [13]. The THPOH finish that has received much commercial attention is THPOH-NH_3. Oxidation with hydrogen peroxide or sodium perborate after curing enhances the resistance of the finish to ultraviolet radiation and weathering but not to sodium hypochlorite. THPOH-NH_3 is used to impart flame-retardant properties to both fabrics and yarns. The synergistic behavior of THPOH finishes has been demonstrated by Horrock [12]. The nitrogen–phosphorus synergism is observed in the phosphonamides but is not present in the simple phosphonate derivatives. Monomeric and dimeric phosphonate derivatives based on *tris*(chloroethyl) phosphonate comprise the Antiblaze (Mobile Chemical Co.) formulations and the precursor for the vinyl phosphonate product, Fyrol 76 (Stauffer). The monomeric and dimeric phosphonates mentioned previously are combined as prepolymers with poly(ethylenamine) in ratios of 2:1 and 3:1, respectively [12]. Curing is done at 160°C, and better crease recovery and abrasion resistance than with the THPOH-NH_3 polymerized products are obtained. This flame-retardant product provides a soft hand to the fabrics, as well as good wrinkle recovery and durable-press properties.

Pyrovatex CP marketed by Ciba-Geigy is a flame-retardant chemical that is obtained from the interaction of dialkylphosphite with acrylamide. A dialkylphosphoric propionamide that interacts with formaldehyde is obtained and yields a methylol derivative [13].

Although other phosphorous-containing derivatives have been developed, their uses are mainly on substrates other than cotton. For example, phosphorous trichloride:*N,N'*-dimethylformamide provides acceptable flame-retardant properties and durability on cotton but deteriorates through ion exchange after repeated launderings. Derivatives of phosphoramide $(P(O){\cdot}NH_2)_3$ are not generally acceptable as flame-retardant finish chemicals since the P–N bond is susceptible to hydrolysis [12]. Tris(1-aziridinyl) phoshine oxide (APO) was used in the past alone with THPC. However, due too its extreme toxicity, APO is no longer commercially available.

Several non-phosphorous-containing flame-retardant chemicals have been described by Horrocks [12]. Included are finishes based on titanium chloride, antimony oxide, DMDHEU, and molybdenum compounds. Combinations of diammonium molybdate with TMM or poly(vinyl bromide) have been used to impart flame-retardant properties on cotton. These chemical combinations also provide resistance to outdoor weathering, which makes the treatments advantageous for tent and sleeping bag fabrics. Another treatment that provides good resistance to outdoor weathering and imparts flame-retardant properties to the fabrics is the use of Sb_2O_3 with halogenated organics. Poly(vinyl chloride)-Sb_2O_3 has also been investigated.

Antimony–halogen synergistic reactions have been described by Pitts [14]. One of the earliest patents citing the use of antimony and halogen in combination for flame-retardant treatment of airplane and balloon fabric was issued in the early 1920s. The most common antimony compounds that have been used for this purpose are the trioxides and trichlorides. Combinations of antimony oxide and brominated aromatic hydrocarbon derivatives applied in the presence of a synthetic resin provide finishes that function well on cotton blend fabrics. The use of synergistic combinations such as these increase the acceptability and usefulness of the finishes by imparting additional qualities that make them beneficial in more end uses. Also, the use of synergistic flame-retardant combinations is important from an economic aspect. The amounts of flame-retardant chemicals used in synergistic reactions are decreased, thereby resulting in a less expensive end product.

D. Flame-Retardant Finishing of Wool

Wool is an extremely flame-retardant fiber due to its composition, which includes 15–16% nitrogen, 3–4% sulfur, and only 6–7% hydrogen and the fact that wool has a high moisture regain at standard conditions. Because wool has high ignition at 570–600°C and flame temperatures of 680°C, it passes most horizontal burning tests. Halo-organic acids and their derivatives, including chloro- and bromo-organic acids were found to provide acceptable flame-retardant properties and durability to laundering and dry cleaning. Both tetrabromophthalic and triboromometanilic acids provide good flame-retardant characteristics to wool but durability of the finishes is poor [12]. The effectiveness of vinyl phosphonates as flame-retardant chemicals for wool has also been examined. Although the vinyl phosphonates provide acceptable flame-retardant properties, the treatments are not durable to laundering.

Titanium (Ti) and zirconium (Zr) compounds have been used successfully as flame-retardant treatments for wool. The treatment (Zirpro), which was developed by the International Wool Secretariat, involves an interaction between wool in the positively charged state and a zirconium or titanium compound such as fluorotitanate as an anionic species. Zirconium is slightly less effective than the Ti but is preferred because it does not cause yellowing of wool as does the Ti.

Because protective clothing may be exposed to a variety of hazards, multipurpose finishes have been used with flame-retardant finishes for wool. Water-, oil-, moth-, and acid-repellent finishes as well as shrink-resistant finishes have been applied with flame-retardant finishes. Organophosphorus compounds such as THPC have been used with other finishes for wool. The application and activity of THPC on wool is similar to that of cotton providing an effective flame-retardant treatment that is durable to at least 50 launderings and to dry cleaning. The compatibility of titanium and zirconium finishes with chemical and resin shrink-resistant treatments has been described by Horrocks [12]. Some multipurpose finishes for wool incorporate a poly(vinylidine chloride) copolymer into a shrink-resist resin such as Synthappret BAP [12]. The Synthappret/Zirpro/fluorocarbon application has been successful, and further work using the Zirpro finish modified with tetrabomomphthalic acid in combination with a neoprene-modified Synthappret BAP formulation was found to provide superior flame-retardant properties and good hand and superior shrinkage resistance [12].

E. Uses of Flame-Retardant Finishes

As mentioned earlier, several problems were noted with fabrics treated with flame-retardant chemicals, specifically, hand of the fabric diminished because many of the treatments increased the stiffness and rigidity of the fabrics. Durability to laundering, dry cleaning, abrasion resistance, and retention of strength of the fabrics also are important parameters that influence the acceptability of fabrics finished with flame-retardant finishes. Given these considerations, there has been a focus on developing and producing flame-retardant finishes that would provide physical and chemical properties desired in fabrics to be used in protective apparel. For example, the U.S. Navy required that uniforms for shipboard personnel be constructed using flame-resistant cotton fabric. The specifications for the uniforms included strength and shrinkage retention. To satisfy these requirements, the USDA Southern Regional Research Center in New Orleans developed flame-retardant finishes that also would provide durable-press properties to cotton fabrics while maintaining the strength specified [14]. A two-step process whereby the flame-retardant chemical, *bis*[tetrakis(hydroxymethyl)phosphonium] sulfate (THPS-NH_3) or precondensate-NH_3, was applied in the first step followed by application of the durable-press treatment (DMDHEU) provided the fabrics with the qualities desired, namely, flame resistance, durable press, strength, and shrinkage retention [14].

Another of the many uses for cotton fabrics treated with flame-retardant chemicals is in protective apparel for foundry workers. The workers are exposed to molten metal hazards and high temperatures from sources of radiant heat such as furnaces and ovens that create heat stress among the exposed workers. Other protective apparel used in this work environment includes aluminum-faced fabric, especially in cases of extreme radiant heat levels, leather (for gloves, safety shoes, etc.), and flame-resistant plastic-coated fabric.

Fabrics constructed of blends of cotton, wool, and polyester are used for protective apparel to impart particular properties such as strength, durability, or dimensional stability. Durable flame-retardant treatments for blended fabrics include *bis*[tetrakis(hydroxymethyl)phosphonium] sulfate (THPS) and trimethylomelamine (TMM), THPC formulations and condensation products of THP salt, and trimethyl phosphoramide (TMPA).

Table 6 Occupations Requiring Protective Clothing

Occupation	Standard
Airline pilots	Air Line Pilots Association–all aircrew uniforms must be flame retardant
Blacksmiths	ANSI/ASC B24.1-1985 Safety Requirements for Forging Machinery
Electrical workers	ASTM F-18—Standard is in development
Fire fighters	NFPA 1971 Standard on Protective Clothing for Structural Fire Fighting: NFPA 1975 Standard on Station/Work Uniforms for Fire fighters
LP gas workers	National Propane Gas Association—Cotton–Polyester must be avoided
Metal industries workers	NFPA-65-1987 Standard for the Processing and Finishing of Aluminum; NFPA 480-1987 Standard for the Storage, Handling and Processing of Magnesium; NFPA 651-1987 Standard for the Manufacture of Aluminum and Magnesium Powder; NFPA 481-1987 Standard for the Production, Processing, Handling and Storage of Titanium; NFPA 482-1987 Standard for the Production, Processing, Handling and Storage of Zirconium
Military personnel	Military-approved clothing must be worn
Petroleum/gas industry	Individual companies have their own FR clothing standards
Steelworkers	ANSI Z229.1-1982 Safety Requirements for Steel Fabrication-Shops Fabricating Steel and Steel Plate
Welders	ANSI/ASC Z49.1-1983 Safety in Welding and Cutting; NFPA 51B 1984 Standard for Fire Prevention in Use of Cutting and Welding Processes
Wildland fire fighters	Forest Service-approved clothing must be worn

Source: Adapted from Ref. 15.

Occupational uniforms for welders, blacksmiths, and gas, metal, and electrical workers frequently are constructed of blended fabrics. Although there are many occupations that are associated with potential fire hazards, relatively few of the occupations have specific standards that require the workers to wear flame-resistant apparel. Table 6 lists those occupations requiring workers to wear protective apparel.

VI. CONCLUSION

Protective apparel used in many occupations is often composed of fabrics treated with flame-retardant finishes. However, inherently flame-resistant textiles are widely used for a multitude of applications. Research in flame-retardant chemicals continues to provide alternatives for those persons requiring garments to protect them from injuries due to fire, heat, chemicals, etc. It is expected that protective apparel will become more specialized in the future [16]. Problems such as comfort, durability, and disposal will be addressed by the new products being developed.

REFERENCES

1. J. Ford, Protective clothing flammability depends on many factors, *Textile Horizons* (May 1989).
2. N. Schultz, *Fire and Flammability Handbook*, Van Nostrand Reinhold Company, New York, 1982.
3. *Annual Book of Standards*, vols. 07.01 & 07.02 (Textiles), American Society for Testing and Materials, Philadelphia, 1991.
4. A. Thornton, Flame retardant chemical market to double in size, *International Dyer & Textile Printer* (1989).
5. Flame retardant buyers' guide, *American Dyestuff Reporter* (1992).
6. W. Baitinger, Flame and thermal resistant workwear fabrics: A product and performance survey, *Industrial Launders Magazine* (1986).
7. N. Hollen et al., *Textiles*, 6th Ed., Macmillan Publishing Company, New York, 1988.
8. B. Smith and I. Block, *Textiles in Perspective*, Prentice-Hall, Englewood Cliffs, N.J., 1984.
9. L. V. McMackin, FP finishing of polyester/cotton fabrics, *Textile Chemist and Colorist* (1977).
10. L. Rubenfeld et al., The thermal and flammability behavior of textile materials, *Applied Fiber Science*, Vol. 2 (F. Happey, ed.), Academic Press, New York, 1979.
11. J. Beninate, Imparting durable press and flame retardancy to cotton fabrics, *Fire Sciences 5* (January/February 1987).

12. A. R. Horrocks, Flame-retardant finishing of textiles, *Rev. Prog. Coloration 16* (1986).
13. M. Lewin, Chemical processing of fibers and fabrics, functional finishes: Part B, *Handbook of Fiber Science and Technology: Volume II*, Marcel Dekker, New York, 1984.
14. J. J. Pitts, Antimony-halogen synergistic reactions in fire retardants, *Flame Retardants, vol. 6, Fire and Flammability Series*, Technomic, Westport, Colo., 1973.
15. C. Coler and J. Colver, Managers, workers must realize need for flame-retardant clothing, *Occupational Health & Safety* (1991).
16. D. Jackson, Niche marketing for protective clothing, *Industrial Fabric Products Review* (1991).

8

Heat- and Fire-Resistant Fibers for Protective Clothing

MASTURA RAHEEL University of Illinois at Urbana-Champaign, Urbana, Illinois

THERESA A. PERENICH The University of Georgia, Athens, Georgia

CHARLES J. KIM University of North Carolina at Greensboro, Greensboro, North Carolina

I. INTRODUCTION

There are three alternative methods for imparting flame resistance to textile goods. These include:

1. Topical treatment in which fabric is treated with a flame-retardant agent
2. Built-in method in which a flame-retardant agent is added to the fiber-forming polymer (spinning dope) in the manufacturing process of man-made fibers
3. Use of inherently heat- and fire-resistant fibers

The first two alternatives have been discussed in Chapter 7. This discussion focuses on inherently heat- and fire-resistant fibers used for protective clothing.

The behavior of human skin exposed to intense heat loads is well documented. The relationship between burn injury threshold and heat flux at the skin surface has been used to assess the performance of protective clothing. The concept of using burn injury threshold for assessing garment protection was first applied by Stoll and Chianta [1]. They derived a relationship between incident heat flux and thermal damage from animal experiments in which pig skin was exposed to fixed levels of heat flux for specific periods. Using their criteria, the projected time to second-degree burn at a heat flux of 330 kW/m^2 is 0.07 sec, whereas at 100 kW/m^2 it is 0.39 sec. Despite the apparent rapid rate of burn development at these intense flux levels, the introduction of a material only 0.5 mm thick increases the protection time at 100 kW/m^2 from 0.39 to 2.5 sec [2], which is adequate for a typical building flashover or mine explosion. The danger lies with parts of the body not covered by clothing, and this is confirmed by statistics that show that some 75% of all fire fighters' burn injuries in the United States are to the hands and face [3].

The main factors that influence burn injury are

1. The incident heat flux intensity and the way it varies during exposure
2. The duration of exposure
3. The total insulation between source and skin, including outerwear, underwear, and the air gaps between them and the skin
4. The extent of degradation of the garment materials during exposure
5. Condensation on the skin of any vapor or pyrolysis products released as the temperature of the fabric rises [4]

One of the most important properties of fabrics intended for use in protective garments is the ability to block heat transfer and reduce burn injury potential. Protection against radiant and convective exposures can be measured in the laboratory using a thermal protective performance (TPP) test. The test uses Stoll criteria to determine a protective index from heat flux measurements [1]. Nevertheless, fabric mechanical integrity in the ensemble also is of great importance. Degradation such as shrinkage, swelling, charring, structural breakdown (fabric cracking open or seam splitting as a result of shrinkage), or the release of pyrolysis products may occur if the heat exposure is sustained or intense.

The most serious garment failure for the wearer is fabric breakdown, seam splitting, or hole formation. When the fabric remains intact, its heat flow properties do not change greatly even when the component fibers are degraded. This is because heat transfer is by conduction and radiation through air in the structure and by conduction through the fibers (which is relatively small). Only when these fibers melt or coalesce and displace the air, or when they bubble and form an insulating char, are heat flow properties substantially altered.

Shrinkage or expansion in the plane of the fabric does not substantially change the thermal insulation of the fabric itself. However, the spacing between fabric and skin or between garment layers may alter, with a consequent change in overall insulation.

Thus, in selection of fabrics for protective clothing against heat hazards, it is not sufficient to rely solely on the heat- and fire-resistance properties of materials used but also the performance of fabric assembly as to mechanical properties, laundry stability, and cost.

A number of heat- and fire-resistant (FR) textile fibers are available, each with its own niche and each with its own set of problems. The following list represents commercial FR fibers.

1. Aramids: Kevlar, Nomex, Twaron, Technora, Conex, Fenilon, Arenka
2. Oxidized acrylic: Panox, Celiox
3. Modacrylics: SEF
4. Novoloide: Kynol
5. Polybenzimidazole: PBI
6. Polyamide-imide: Kermel
7. Polyphenylene Sulfide: PPS
8. Carbon fiber: rayon-based, PAN- (polyacrylonitrile) based, pitch-based
9. Glass: Zetex
10. Steel
11. Asbestos
12. Potytetrafluoreoethylene: PTFE
13. Polyvinyl Chloride: PVC
14. FR Modacrylic: Teklan
15. FR Polyester
16. FR Viscose: Asgard
17. Zirpro Wool
18. FR-treated cotton

This chapter will focus on inherently heat- and fire-resistant fibers of commercial significance for protective clothing applications.

II. HEAT- AND FIRE-RESISTANT FIBERS

A. Aramid Fibers

Currently on the world market aramid fibers include DuPont Company's Nomex and Kevlar; Twaron of Akzo NA (Netherlands); Technora and Conex of Teijin Ltd., Japan; and Apyeil of Unitika, Japan. Aramids are aromatic polyamides that have unique properties compared to aliphatic polyamides such as conventional nylons. Aramid fibers were developed primarily for their inherent heat and fire resistance and high strength, particularly those of Kevlar, Twaron, and Technora.

Nomex, Conex, and Apyeil are meta-oriented fibers, that is, they are composed of polymetaphenylene isophthalamide (MPD-I). The polymer structure is as follows:

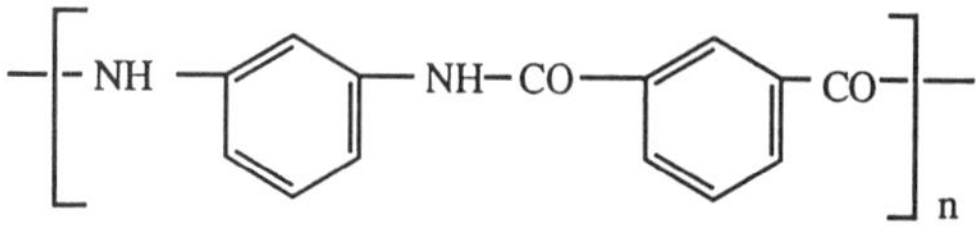

MPD - I

The polymer containing meta-oriented phenylene groups exists in folded chain configuration and, hence, exhibits tensile properties in the range of conventional polyamides (nylon), but with exceptional fire resistance.

Kevlar, Twaron, and Technora are parapolyphenylene terephthalamide (PPD-T) fibers. The para orientation of phenylene groups in these polymers makes a significant difference in melt temperatures—424°C for Nomex versus 503°C for Kevlar. Also, the tensile properties of PPD-T fibers are very high, that is, these fibers are ultra-high-strength, high-modulus fibers. However, heat- and flame-resistant aramid fibers contain a high proportion of meta-oriented phenylene rings.

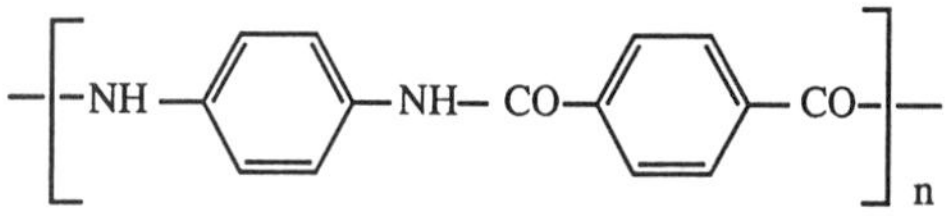

PPD - T

Aromatic polyamides are also distinguished from aliphatic polyamides (nylons) by their excellent resistance to corrosive atmospheres, flame resistance, and resistance to degradation by high-energy particles and gamma ray radiation. Also, fibers retain high strength, good recovery, and flex life at elevated temperatures and are readily crystallizable.

The classical synthesis of aromatic polyamides was achieved by Kwolek, Morgan, and others [5–9] who investigated many compositions. Blades [10] succeeded in producing high-strength fibers by optimizing polymer solution and spinning. Ozawa et al. [11] further broadened the polyamide compositions by incorporating pivotal groups [12]. Nomex was marketed by DuPont Company in 1961, whereas Kevlar was commercialized in 1972. Twaron is similar to Kevlar and was introduced by Akzo NA in 1987.

Technora was introduced by Teijin Ltd. in 1985. The fiber differs from Kevlar in its polymer composition and method of preparation. It is a wholly aromatic copolyamide, highly modified with a crankshaft-shaped coingredient [12]. It has modest crystallinity, but a fully extended chain structure. Hence, it offers some unique fiber properties such as chemical and impact

Table 1 Properties of Commercial Aramid Fibers

Fiber	Density, g/cc	Tenacity, g/tex	Extension, %	Initial modulus, kg/mm	Thermal decomposition temperature °C	Limiting oxygen index, %
Nomex	1.38	54	15	1,850	424	26
Kevlar	1.44	210	3.6	6,000	—	24.5
Kevlar 49	1.44	200	2.5	8,500	503	24.5
Twaron						
Technora	1.39	225	4.4	7,100	500	25
Fenilon	—	45	15–20	—	—	—
Terlon	—	80–100	1–1.5	8,500	—	—
Vinivlon	—	150–170	3–7	12,000	—	—

resistance, in addition to high strength, modulus, and heat and fire resistance. The basic properties of these fibers are presented in Table 1.

1. Heat and Fire Resistance

Aramid fibers offer excellent thermal stability. The fibers do not melt or drip with heat but decompose at their melting temperature in the range of 350–550°C. The fibers have low shrinkage at high temperatures in the order of -5.7×10^{-6}/°C in the temperature range of 20–80°C. However, the shrinkage trend is nonlinear in the range of 20–400°C. Nomex has a higher thermal shrinkage rate than Kevlar.

In protective clothing it is important that the ensemble not shrink and the fabric retain its strength after exposure to high temperature so that the garment does not break open under stress. The tensile properties of important commercial aramid fibers at elevated temperature (250°C) are equal to those of conventional textile fibers at room temperature [13]. The tensile moduli of aramid fibers are somewhat lower at 300–350°C, and elongation is lower at 300°C than at room temperature. Table 2 summarizes the effect of elevated temperature on fiber properties.

Long-term performance of aramids at elevated temperatures is predicted from tensile properties after heat aging in air. The effect of heat aging at 300°C in air is shown in Table 2. Aramids retain useful tensile properties for 1–2 weeks even under these severe conditions. However, performance decreases with increasing temperature and hours of exposure. Both tenacity and elongation are reduced with heat aging. However, ultimate failure of the fiber is due to loss of elongation rather than loss of tenacity.

Table 2 Effect of Heat on Fiber Properties

	Nomex			Kevlar		
Density, g/cc	1.38			1.44		
Tenacity, N/tex	0.49			0.56		
Elongation to break, %	28			5.7–20		
Initial modulus, N/tex	10.0			6.2		
Moisture regain, %	3.5			7.0		
Temperature to degrade, °C	440			550		
	Long-term performance after 300°C exposure Days					
	1	7	14	1	7	14
Tenacity	0.26	0.15	0.063	0.40	0.12	0.20
Elongation	9.9	5.7	16	16	—	6.9
Initial modulus	8.6	6.8	41	5.9	—	21

Aramid fibers characteristically burn only with difficulty and degrade rather than melt. Their limiting oxygen index (LOI) values, generally accepted as a measure of flame resistance, are high. For example, it is 26.0 for Nomex and 24.5 for Kevlar. The LOI values of Nomex and Kevlar can be increased by incorporated tetrakis(hydroxymethyl)phosphonium chloride (THPC) and subsequent crosslinking reaction. The LOI values of 28–30 can be achieved for Kevlar and 40–42 for Nomex [13].

2. *Resistance to Chemicals*

The resistance of aramid fibers to chemicals is good but not outstanding. These fibers are much more resistant to acids than nylon, but not as acid resistant as polyester fibers, except at elevated temperatures. Resistance to a strong base is comparable to that of nylon 6-6 fibers.

Table 3 summarizes the chemical resistance of aramid fibers.

3. *Resistance to Sunlight and UV Radiation*

Protective heat-resistant clothing may be exposed to sunlight and ultraviolet (UV) radiation in the course of work. Aramid fibers, like the nylons, are susceptible to UV degradation. Photoquenchers that stabilize nylon do not have the same effect on aramids and are not effective.

Table 3 Chemical Resistance of Aramid Fibers

Chemical	Concentration, %	Temperature, °C	Time, hr	Effect on breaking strength[a]	
				Nomex	Kevlar
Acids					
Formic	90	21	10	None	None
Hydrochloric	10	95	8	None	Degraded
	35	21	10	Appreciable	Degraded
	35	21	100	Appreciable	Degraded
Sulfuric	10	21	100	None	Slight
	10	60	1,000	Moderate	Appreciable
	70	21	100	None	Slight
	70	95	8	Appreciable	Degraded
Alkalies					
Ammonium hydroxide	28	21	100	None	None
Sodium hydroxide	10	95	8	Appreciable	Degraded
	40	21	10	None	None
Other chemicals					
Dimethylformamide	100	70	168	None	None
Perchloroethyene	100	70	168	None	None
Phenol	100	21	10	None	None
Sodium chlorite	0.5	21	10	None	Slight
	0.5	60	100	Moderate	Slight

[a]Strength loss: none, 0–9%; slight, 10–24%; moderate, 25–44%; appreciable, 45–79%; degraded, 80–100%.

4. *Applications of Aramid Fibers*

High-temperature and fire-resistant fibers are needed for many industrial, military, and aerospace applications for extended periods of time at temperatures of 200–400°C. In the end uses such as fire fighting, more stringent heat and fire resistance is required, Nomex and Apyeil exhibit good flame resistance but may not provide complete flame protection in catastrophe such as total immersion in flame. Kynol novalac fiber and polybenzimidazole (PBI) fibers are known to offer improved flame resistance under these conditions. Nevertheless, aramid fibers and blends of aramids with carbon or glass fibers are effective as protective garments for smelting and foundry operations. Where metalized heat protective clothing is needed, aramids give successful protection against splashes of molten iron and steel up to about 1600°C. However, direct temperature exposure should not exceed 400°C [14]. Aramids are good contenders as substitutes for asbestos heat protective clothing because of their better textile properties [14]. Although high-tenacity Kevlar fibers are heat resistant and have a high melting point, which is > 500°C with decomposition, they are not particularly useful as heat-resistant fibers. The fibers have low elongation and lose elongation rapidly at elevated temperatures, resulting in brittleness. However, Kevlar fibers have high shrink resistance and when phosphorous compounds are incorporated, these fibers have excellent flame resistance. Hence, these fibers in blends with Nomex are often used in heat- and flame-resistant garments [13].

B. Polybenzimidazole Fibers

An acronym for poly[2,2′-(*m*-phenylene)-5,5′-bibenzimidazole] is PBI. It is marketed as a fiber by Celanese Corporation. Its chemical structure is

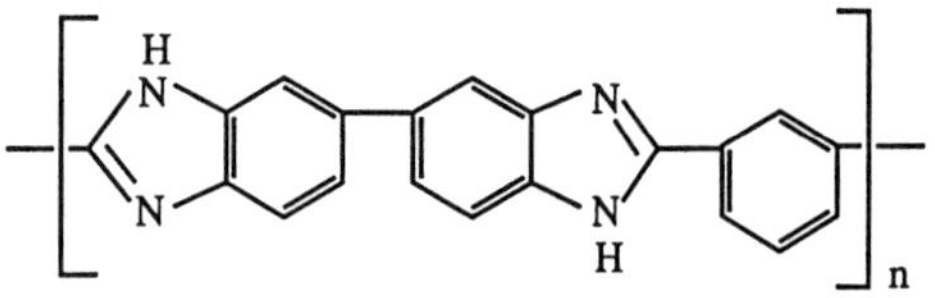

PBI

The fiber combines resistance to high temperatures and chemicals with excellent textile and tactile performance.

Polymers containing stable imidazole links were first prepared successfully in 1959 by Brinker and Robinson [15]. The compositions were based on bis-*o*-diaminobiphenyls and aliphatic dicarboxylic acids. A modification of these combinations was used by Vogel and Marvel [16,17] to prepare wholly aromatic polybenzimidazoles, and these were shown to have melting points and thermal stabilities considerably higher than those of the aliphatic analogs.

Originally, the developmental work in the area of PBI textile fibers was done at DuPont for the U.S. Air Force. The Air Force was interested in the development of high-temperature-resistant webbing and drogue chutes for deaccelerating jets on landing. Limited quantities of the fiber were successfully dry spun at DuPont [18]. The fibers were strong and thermally stable. In 1963 an Air Force contract was issued to Celanese Research Company for production of fiber. After full-scale evaluation by the U.S. Air Force, it became apparent that PBI not only was a very comfortable fiber for wearing apparel but had a very high degree of heat and flame resistance [18]. In conventional flammability tests, PBI does not burn in air, and its limiting oxygen index is 41%, which is higher than aramid fibers. This observation led to the design of PBI-based protective clothing for pilots and other crew members and for personnel in areas where fuel fires were a hazard.

Field evaluations of flight suits of various materials were carried out jointly by the U.S. Air Force Materials Laboratory and the U.S. Army Natick R & D Laboratory (Mass.). The results showed that PBI offered superior thermal and fire protection over other materials similarly tested. They also showed that shrinking of the fabric occurred at temperatures above the glass-transition point, which would have exposed parts of the body of the wearer to the flame [19]. Treatment of PBI with chlorosulfonic acid in phosphorus oxychloride followed by heat treatment raised the oxygen concentration required to sustain combustion and reduced the flame shrinkage significantly. An alternative approach to thermal stabilization applied a sulfuric acid treatment followed by heating. A two-stage sulfonation process reduced the flame shrinkage from 50% to 6%. Flammability was not affected, and the fiber showed improved moisture regain, acceptable tensile properties, and better resistance to laundering versus the unstabilized product [19]. Commercial production of PBI fiber by Celanese Corporation began in 1983.

1. Textile Properties

The textile properties of stabilized and unstabilized fibers are given in Table 4. Only the stabilized fiber is now sold by the Celanese Fibers Marketing Corporation. An increase in fiber density occurs during the stabilization treatment as a result of the incorporation of 5–6% sulfur. It accounts for the changes in tensile strength and modulus, which are expressed as N/tex.

Table 4 Textile Properties of PBI Fibers

Properties	Stabilized	Unstabilized
Filament fineness, tex	0.16	0.16
Tenacity, N/tex	2.3	2.7
Breaking elongation, %	30	30
Initial modulus, N/tex	39.6	79.2
Density, g/cm	1.43	1.39
Moisture regain at 20°C, 65% rh, %	15	13
Shrinkage in air at 204°C, %	<1	>2
Flame-test shrinkage, %	6	50
Shrinkage at 400–500	4	10
Limiting oxygen index, %	41	—

Source: Adapted from Ref. 19.

2. Flammability

PBI fiber has a low rate of gas evolution in air, and the off gases are relatively inert. Thus, PBI is intrinsically nonflammable in air [20]. When the fabric strip is burned at the top, the LOI is 46%, whereas for burning from the bottom upward is 28%. The temperature at which burning is sustained at 46% oxygen is 720°C for the surface and 1180°C in the flame. Stabilization with sulfuric acid and heat improves the dimensional stability from 50% to 6% shrinkage in flame tests.

3. Properties at Elevated Temperatures

As with other fibers, tenacity changes with temperature; however, elongation of the fiber remains essentially constant up to 350–370°C and then drops off fairly rapidly from 30% to reach about 5% at 450°C within 2 min. Initial modulus remains constant over the temperature range and starts to increase near 500°C as a result of oxidative crosslinking.

The useful life of PBI fibers at elevated temperatures depends on the amount of oxygen in the environment. In inert atmosphere or vacuum, no significant aging effects have been observed, even at 350°C after 300 hr. Nomex and Kevlar lose approximately 50 and 86%, respectively, of their initial strengths at 300°C after 300 hr [19].

4. Chemical Stability

Stabilized PBI fiber exhibits outstanding stability in hostile environments [18]. It is hydrolytically stable and retains 96% of its strength after exposure to 149°C for 72 hr in a 67-psi steam atmosphere and loses little or no strength after 16 hr at 182°C and 140-psi steam [18]. Its resistance to degradation when exposed to concentrated acids and bases is outstanding—that is, much greater than Nomex and Kevlar. Results of strength retention after exposure to acids and bases are summarized in Table 5.

PBI also is resistant to organic chemicals. Fiber strength is unaffected by a wide range of organic liquids such as acetic acid, methanol, perchloroethylene, dimethylacetamide, dimethylformamide, dimethylsulfoxide, acetone, kerosene, and gasoline [21].

5. Wear Comfort

PBI fibers exhibit a high moisture regain (15%), which is enhanced by the acid-stabilization process. This high moisture regain, together with the soft hand of PBI, contributes to wear comfort. As part of a large comfort development program, PBI was evaluated against competitive fibers by the Gillette Research Institute. In the study, wearers were asked to evaluate the garments

Table 5 Strength Retention after Immersion in Inorganic Acids and Bases

	Concentration, %	Temperature, °C	Time, hr	Strength retention, %
Acids				
Sulfuric	50	30	144	90
Hydrochloric	35	30	144	95
Nitric	70	30	144	100
Bases				
Sodium	10	30	144	95
Potassium	10	25	24	88

Source: Adapted from Ref. 21.

on a five-point scale (5 = total comfort) for various descriptors. The test subjects rated the garments' comfort while exercising (Exercycle, 10 min at 13–20 km/hr) and while at rest under various environmental conditions (17–35°C, 20–75% relative humidity). The garments were laundered (10 min at 49°C, tumble dried) twice before each exposure condition. In the study, PBI was rated as comfortable as cotton and rayon [22]. In military evaluations of flight suits, PBI was the preferred fiber for comfort [23–25]. It has also been proposed that the high moisture regain of PBI contributes to improved thermal protective performance, that is, PBI fabrics require more heat to dissipate the additional moisture [26].

6. *Applications*

First used in various National Aeronautics and Space Administration (NASA) applications, PBI has now successfully entered commercial uses such as protective apparel. Protective gear for fire fighters uses the nonflammability of PBI and its ability to form a flexible, nonbrittle, nonshrinking char that preserves the protection after flame exposure [21]. An outer-shell fabric for coats is based on a 40% PBI:60% high-modulus (HM) aramid fabric, which exceeds the 1971 National Fire Protection Association (NFPA) requirements [19]. Protective hoods and helmet liners of 20% PBI:80% flame-resistant rayon are used to provide flame and thermal protection for fire fighters and racing car drivers, and other work environments where facial flame protection is required and comfort is a critical requirement.

Heat protective gloves for metal and glass industries traditionally are made of asbestos. These gloves are often subjected to severe abrasion and short-duration temperatures up to 815°C. Because of these harsh conditions, asbestos gloves deteriorate rapidly. Gloves containing PBI fiber can stand up to more exposures than gloves containing asbestos. Gloves properly engineered using PBI fiber are softer and more supple, offering the worker greater mobility, comfort, and protection, even if the fabric becomes charred [27]. Also, field trials have demonstrated the cost-effectiveness of gloves made from a PBI/high-strength aramid/glass fabric blend in comparison with asbestos gloves. Although initially more expensive, PBI-blend gloves lasted up to eight times longer than asbestos gloves [27]. In foundry applications, workers reported longer exposure times with light-weight (270-g/m^2) aluminized PBI/FR rayon fabrics than with conventional aluminized glass fabrics (640 g/m^2). In evaluations for molten metal splash, aluminized PBI fabrics offer better thermal protection than heavier aluminized asbestos fabrics [28–30].

Thus, the development of PBI fiber is a significant step forward in improved heat- and fire-resistant protective clothing. Originally designed for NASA applications, PBI is now available for various commercial protective apparel end uses. The improved flammability, thermal and chemical resistance, comfort, and textile processability of PBI provide new levels of safety for industrial workers.

C. Novoloid Fibers

Novoloid fibers, also known under the trade name Kynol, are crosslinked phenolaldehyde fibers prepared by fiberizing and curing novolac resin [31]. In contrast to the linear, crystalline structure of most synthetic organic fibers novoloid fibers have an amorphous, three-dimensionally crosslinked ''network'' polymer structure. A typical segment of this structure is

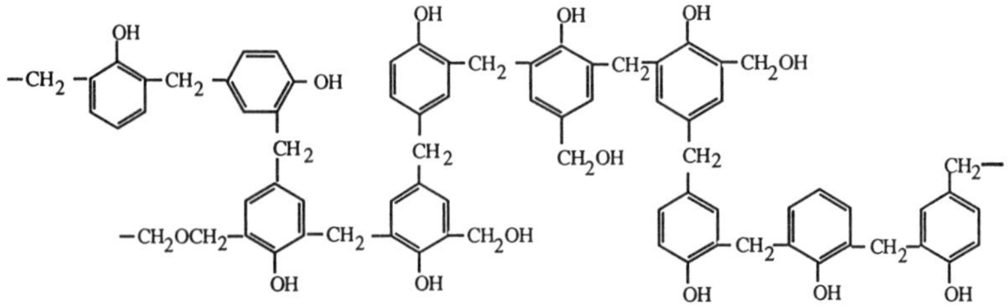

Novoloid

The fibers are soft to the touch and are light gold in color, darkening with age or exposure to light without significant change in properties. In strength they are roughly comparable to the man-made cellulosics (rayons). Typical textile properties are summarized in Table 6. The ranges shown for strength, elongation, and modulus reflect the dependence of these properties on fiber diameter, with the higher values corresponding to finer diameters.

Table 6 Textile Properties of Novoloid Fibers

Property	Value
Diameter, μm	14–33
Specific gravity	1.27
Tensile strength, mN/tex	120–160
Elongation, %	30–60
Modulus, mN/tex	260–360
Elastic recovery, %	92–96
Work to break, mN/tex	26–53
Moisture regain at 20°C, 65% rh, %	6

Novoloid fibers are processed by suitably modified conventional textile techniques. The moderate tensile strength (comparable to that of cellulose acetate) and lack of inherent crimp require careful processing procedures to prevent excessive fiber breakage. Blending with other fibers such as aramids improves processing speeds and increases yarn tensile strength. Heavy yarns (300 tex) of 100% novoloid fiber are readily spun. With suitably modified equipment, 100% novoloid yarns as fine as 30 tex and blended yarns of 20 tex are routinely produced.

Spun novoloid yarns are used in the production of woven fabrics in weights from 100 to >550 g/m^2, as well as in knitted products such as gloves. Blending with other fibers improves the low tensile strength and abrasion resistance of 100% novoloid materials. Novoloid fibers may be dyed with cationic or disperse dyes; however, color range and stability are limited by the inherent gold color of the fiber and its tendency to darken with exposure to heat and light.

1. *Heat Resistance*

Novoloid fibers are highly flame resistant but are not high temperature fibers in the usual sense of the term. A 290-g/m^2 woven fabric withstands an oxy-acetylene flame at 2500°C for 12 sec or more without breakthrough. However, the practical temperature limits for long-term application are 150°C in air and 200–250°C in the absence of oxygen.

2. *Fire Resistance*

With a limiting oxygen index of 30–34, novoloid fibers will not support combustion in the atmosphere. Moreover, when exposed to flame, novoloid fibers do not melt but gradually char until completely carbonized without losing their initial fiber form and configuration. This behavior is attributable to the crosslinked, amorphous, infusible structure of the fiber and to its high (76%) carbon content. Above 250°C in the absence of oxygen, novoloid fibers undergo gradual weight loss until, close to 700°C, the fiber is fully carbonized, with a carbon yield of 55–60% [32,33]. Thus, the high initial carbon content results in limited production of volatiles (weight loss 40–45%), only a portion of which is, in fact, flammable. Melting does not occur, and shrinkage is small, suggesting that the crosslinked structure of the material promotes gradual coalescence of the aromatic units into a stable, amorphous char; the relatively low thermal conductivity of the material aids in the moderation of this process.

Exposure to flame results in formation of an amorphous surface char that serves both to radiate heat from the fiber and to retard the evolution of flammable volatiles; the latter is limited by the high carbon content. The amorphous nature of the char presents a minimum reactive surface to the flame. Moreover, in a textile-fabric structure, the charred fibers at the fabric face provide a protective insulating barrier retarding penetration of both heat and oxygen into the interior of the fabric; minimal shrinkage enhances the mechanical stability of this barrier. Finally, the formation of H_2O and CO_2 as products of decomposition and combustion provides an ablative type of cooling effect.

The rate of carbonization depends on temperature, time, and presence or absence of oxygen. Weight loss accompanying carbonization is detectable in thermogravimetric analysis (TGA) at about 300°C and continues to about 700°C, with a final carbon yield of 55–60 wt% (94 to 95% carbon). The loss of weight is accompanied by some loss of strength and elongation, as well as approximately 20% shrinkage. In addition, densely packed and poorly ventilated masses of novoloid fibers, when heated in air, may be subject to exothermic decomposition through "punking." Thus, as a practical matter, and despite their excellent resistance to flame, temperature limits for long-term applications of novoloid materials are 200–250°C in the absence of oxygen and 150°C in air.

Because novoloid fibers are composed only of carbon, hydrogen, and oxygen, the products of combustion are principally water vapor, carbon dioxide, and carbon char. Moderate amounts of carbon monoxide may be produced under certain conditions; but the HCN, HC1, and other toxic by-products typical of combustion of many flame-resistant organic fibers are absent. The toxicity of the combustion products is thus very low or negligible. [34]. Smoke emission is also minimal, less than that of virtually any other organic fiber.

3. *Chemical Resistance*

Novoloid fibers exhibit excellent chemical and solvent resistance, as shown in Table 7. They are attacked by hot or concentrated sulfuric and nitric acids and by strong bases but are virtually unaffected by nonoxidizing and organic acids, including hydrofluoric and phosphoric acids, dilute bases, and organic solvents [35].

4. *Protective Clothing Applications*

Protective apparel for fire fighters, racing car drivers, and others similarly at risk to accidental exposure to intense flame must have the combined charac-

Table 7 Chemical and Solvent Resistance of Novoloid Fibers

	Concentration, wt %	Temperatures, °C	Time, hr	Percent strength loss			
				<10	<25	<45	<80
Strong mineral acids							
Hydrochloric	35	98	1000	x			
Hydrochloric	50	60	1000	x			
Sulfuric	10	98	1000	x			
Sulfuric	60	60	100		x		
Nitric	10	20	100	x			
Nitric	10	98	8				x
Phosphoric	85	135	1000	x			
Fluoric	15	50	40	x			
Organic acids							
Acetic	100	98	100	x			
Formic	91	93	100	x			
Oxalic	10	98	100	x			
Strong alkalis							
NH_4OH	28	20	100	x			
NaOH	10	60	100	x			
NaOH	40	98	8		x		
Solvents							
Benzene	100	80	1000	x			
Acetone	100	56	1000	x			
Gasoline	100	20	1000	x			
Gas							
Steam	100	155	100	x			

Source: Adapted from Ref. 35.

teristics of flame resistance, insulation, and wearability. A study performed by the U.S. Air Force on heat transfer characteristics of flight jacket materials reached the conclusion that a 136-g/m^2 (4-oz/yd^2) novoloid batting, used in conjunction with an aramid outer shell fabric, provided the best protection from fire; it was superior to all other combinations evaluated in retarding heat transfer and thus preventing burns [36]. Novoloid fibers have excellent fire-blocking properties but lack durability; hence, they find applications in aircraft interiors as a fire-blocking layer between the seat cushion and upholstery material that meets the FAA regulations [37]. In addition, novoloid fibers are used as precursors for carbon or activated carbon fibers. These activated carbon fibers offer great potential for use in the medical field as wound dressings and as purifying filter materials.

D. Polyphenylene Sulfide Fibers

Polyphenylene sulfide (PPS) is formed by the reaction of *p*-dichlorobenzene and sodium sulfide. Since 1973, the polymer has been marketed by Phillips Petroleum and sold under the name of Ryton [38]. The chemical structure of PPS described by Scruggs [39] follows.

$$-\left[C_6H_4 - S \right]_n -$$

PPS

The process of producing PPS was described by Edmonds and Hill in U.S. Patent #3,354,129 (to Phillips Petroleum) in 1967. In the process, para dichlorobenzene and sodium sulfide react in a polar solvent at elevated temperatures forming a linear, *p*-linked aromatic sulfide polymer [39]. Polyphenylene sulfide resins are characterized by useful properties, including good thermal and mechanical stability, excellent resistance to chemicals and environmental conditions, and nonburning and nondripping flammability behavior [38].

Polyphenylene sulfide resins were prepared as early as 1888 by researchers studying aromatic sulfur compounds. However, not until 1948 did PPS resins begin to be prepared with accuracy and precision [40]. Macallum, considered the founder of polyphenylene sulfide, investigated PPS by reacting *p*-dichlorobenzene with sulfur and anhydrous sodium carbonate in a sealed vessel at 340°C [41]. In 1954, Dow Chemical Company purchased the patents from Macallum to further their research involving thermally stable plastics. The use of PPS by Dow was limited to injection molding and coatings [42].

In 1967, Edmonds and Hill sold a patent to Phillips Petroleum Company for a new synthesis involving polyhalo-substituted aromatic compounds [42]. Phillips was not only interested in PPS for its use in injection molding but also for its capabilities in the fiber market. Before PPS could be used as a fiber, it would need to be suitable for melt spinning. Because neither of its current forms, the low molecular weight version nor the thermally cured higher molecular weight version, could be used in the melt spinning process, modifications were needed [39]. The modifications included (1) implementing a polar solvent containing alkali metal carboxylate, (2) dehydrating the sulfide stream, (3) polymerizing with dichlorobenzene, (4) polymer recovery, (5) polymer washing, (6) polymer drying, and (7) packaging [38]. After the process was perfected, the resulting fiber was marketed under the name Ryton [38].

1. *Textile Properties*

The textile properties of staple and monofilament fibers are given in Table 8. For staple PPS fibers, different processing conditions can create certain desirable end-use characteristics. For example, boiling water shrinkage can be either low (0–5%) or high (15–25%) depending on the processing methods used.

Table 8 Textile Properties of PPS Fibers

Properties	Value
PPS stable fibers	
Denier per filament	3
Tenacity, gpd	3.0–3.5
Elongation %	25–35
Initial modulus, gpd	30–40
Moisture regain %	0.6
Melting point, °C	285
PPS monofilament fibers	
Denier	350–400
Tenacity, gpd	3–4
Elongation, %	12–16
Initial modulus, gpd	45–55

Source: Adapted from Ref. 39.

2. *Thermal Stability*

The temperature stability of PPS is very good in all areas of thermal testing—flammability testing, limiting oxygen index, thermogravimetric analysis, and long-term exposure to high temperature. When a sample of PPS is exposed to fire, it will continue to burn until the externally applied flame is removed. After the flame is removed, the sample will stop burning almost immediately [39]. According to the National Bureau of Standards, PPS maintains a low fame spread rating along with a low smoke density rating.

The concentration of oxygen required for a material to burn, or limiting oxygen index for PPS fibers, is approximately 35 [39], which classifies PPS as nonflammable.

Thermogravimetric analysis measures the weight loss of a specimen after that specimen is exposed to increased temperatures. For PPS, a rapid weight loss occurs at about 500°C. The weight remains almost constant at about a 40% weight retention up to about 1000°C [39].

An Instron environmental test chamber was used to measure PPS for retention of tensile properties at elevated temperatures. It was found that PPS retained approximately 60% of its tensile strength at 200°C, and after exposure to a 250°C environment, PPS retained approximately 40% of its tensile strength [39].

After testing 12 different types of plastics, NASA reported that PPS possessed the highest degree of fire safety. The composition of the products produced from the material in a fire is an important criterion in determining the safety of a material. Pyrolysis of PPS polymer in air at 700°C produced mainly carbon dioxide (CO_2), sulfur dioxide (SO_2), and carbon monoxide (CO) [39].

3. *Chemical Stability*

Scruggs and Reed [39] indicated that PPS fiber retains its properties under extremely adverse conditions. The chemical resistance of PPS is excellent and is considered second only to fibers made from polytetrafluoroethylene [39].

The PPS fiber exhibits good resistance to both organic and inorganic chemicals. PPS has excellent resistance to hydrochloric and sulfuric acids, carbon tetrachloride, chloroform, sodium hydroxide, formic acid, perchloroethylene, and acetic acid. However, its resistance to concentrated nitric acid and 50% chromic acid is very poor. The only organic chemical that affected PPS negatively was toluene, which produced a maximum strength loss of 25%. Polyphenylene sulfide generally is unaffected by most organic and inorganic acids, aqueous bases, specific amines, hydrocarbons, and most chlorinated organics [43].

4. Applications

Polyphenylene sulfide fibers have properties that make them desirable not only in high-temperature situations but also in situations involving chemical contamination. For example, a specific area in which PPS fibers are used is in the petroleum industry where protective clothing is used by workers cleaning up spills [40].

Currently, PPS is being used in aircraft interiors because of its excellent resistance to flammability. In fact, NASA promotes the use of PPS in all industrial aircraft, except in specific and atypical circumstances [39].

However, the primary use of PPS is in filtration media. The PPS fiber is used as a filter fabric for baghouses for coal-fired boilers. PPS is particularly suited for this application, because high flue gas temperatures make it difficult to filter fly ash from coal-fired boilers in industrial and utility plants. In this application, fabrics that are capable of long-term exposure to acidity, high temperatures, and abrasion are needed, and fabrics made of PPS fibers are well-suited in this application [39].

Other potential applications of PPS fiber include needlepunch felt and belts used in the paper industry in its pressing and drying operations; specialty papers for electrical applications; specialty filters that could be used for filtering hot, corrosive chemicals; and woven fabrics made from monofilament and continuous filament yarns that could be used as demisters where chemical resistance is essential [39].

E. Glass Fibers

Glass fibers are created from mixtures of silicates and borosilicates in the form of mixed sodium, potassium, calcium, magnesium, aluminum, and other salts [44]. The chemical structure of glass fiber follows:

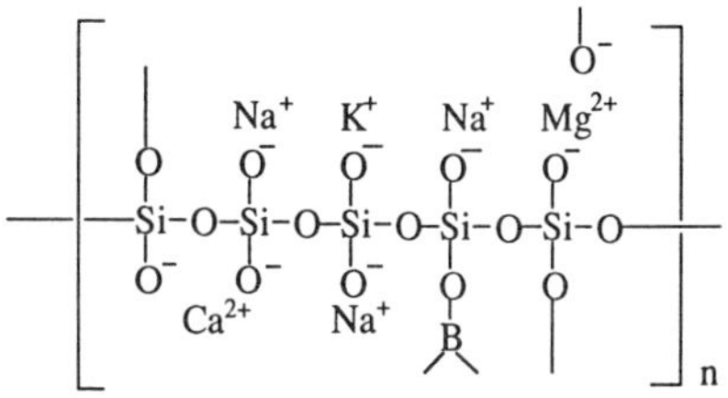

Glass

Glass fibers contain many of the same characteristics that are associated with bulk glass and therefore do not burn, shrink, stretch, mildew, or rot. However, glass fibers are flexible enough to be used in apparel where heat resistance is of primary importance [45].

Glass fibers were first obtained in experiments as early as the mid-1600s by physicist Robert Hooke. But, glass fibers that could be woven into fabric did not exist until 1893, when Edward D. Libby, founder of Libby Glass Company, made filaments that could be woven into fabric. Glass fibers first were used semiregularly in World War I by the Germans for thermal insulation. The next largest advance in the use of glass fibers was in 1931 when Owens Illinois Glass Company used a steam-blowing technique to produce glass fiber. In 1938, Owens merged with Corning Glass Works to form the Owens–Corning Fiberglas Corporation, which is still in existence today [45]. The consumption of glass fibers in the United States grew from 23.5 million pounds in 1950 to 282.6 million pounds in 1965 [46]. Composites made of glass fibers have been used extensively in the last 20 years for industrial, military, and consumer uses [46].

1. *Physical Properties*

The dimensions and arrangement of ions in the silicate network impact on the properties of different glasses. Oxygen and cations of silicon, boron, aluminum, calcium, magnesium, lead, barium, zinc, sodium, and potassium in varying combination form the basis for glass fibers [47]. A basic characteristic of glass structures is that the network in silicate glasses is irregular and aperiodic.

The physical properties of glass are extremely important due to the fact that glass dramatically decreases in tensile strength when there are defects in the inner fiber construction [47]. The defects usually result from lack of quality control and inadequate standards during the processing stage. A consistent drawing process is essential to obtain accurate results with the formation of glass fibers, and the environment must be free of dust and other particles that may interfere with the post-molten stage [47].

Glass has a specific strength of 0.60 gpd for the filament form and up to 8.05 gpd for the staple form [48]. In general, glass fibers are two times as strong as the strongest textile fibers on an equal weight basis. For example, the tensile strength of nylon is 6.0 gpd compared with 15.6 gpd for glass [48].

Glass fibers have a tenacity of 15.3 gpd compared to nylon with a tenacity of 6.0 gpd and polyester with 5.2 gpd. In addition, glass fibers have an elongation of 4.8%, which makes the fibers extremely dimensionally stable. Glass fibers do not shrink, stretch, or deform easily [48]. Fabrics made from glass fibers have excellent wrinkle resistance and recoverability when they have been heat set properly. Because glass fibers are quite stiff and brittle and break readily on bending, they show poor abrasion resistance [44].

2. *Chemical Properties*

Glass fibers are chemically stable under all but very extreme conditions. Under normal circumstances, oxidation, reduction, and biological agents do not affect glass fibers [44]. Glass has excellent resistance to sunlight, weather, fungi, microorganisms, oils, solvents, weak alkalis, and corrosive vapors [45]. However, hydrofluoric acid and concentrated alkalies will deteriorate glass fibers when exposed to these chemicals over extended periods of time [44].

3. *Thermal Properties*

The thermal stability of glass fibers is superior. Glass fibers are resistant to burning and provide excellent protection against high temperatures. Glass fibers remain dimensionally stable even in contact with flame. Glass fibers retain 50% of their tensile strength when confronted with temperatures of 700°F and 33%, with temperatures of 1000°F.

4. *Applications*

The excellent thermal and chemical characteristics of glass fibers make them amenable for use in many industrial and commercial applications. One of the most widely used applications of glass fibers is in industrial filtration. The resistance of glass fibers to thermal decomposition makes them ideal in glass filtration fabrics, which when properly constructed and finished are 99.4% efficient on particles of less than micron size and are important in air pollution control [48].

Glass fibers have been used successfully in high-temperature filtration situations in industries such as steel, fertilizer, gray iron, cement, nonferrous metal, and chemical industries. Glass fabric has the advantage of providing extremely high collection efficiencies at elevated temperatures on a reliable basis [48].

In the electrical industry, glass fiber has been recognized as an ideal supporting and reinforcing base for dielectric materials. The properties of glass fiber such as high thermal endurance, high tensile strength, superior electrical properties, chemical resistance, and weathering resistance are the focus of its use in the electrical insulation industry [48].

Beta glass fibers have been used in protective clothing and accessories. However, the very high cost of beta fibers has made it difficult to market the fibers widely.

Recently, yarns made of 99.8% pure silica were introduced [49]. Applications for the silica yarns include high-temperature mats, electrical boards that require resistance to high temperature, and high-temperature sewing thread.

III. PROTECTIVE PERFORMANCE OF BLEND FABRICS

Although a wide range of heat- and fire-resistant fibers is available for specific end uses, there is no all-purpose fiber for heat and fire protective clothing. The blending of fibers with complementary properties is one of the solutions. Others are the use of novel yarn and fabric structures, chemical treatments, and coatings.

Protective clothing is such a broad area that one fiber or one type of blend cannot cover all requirements. Heat and fire protective clothing will vary in critical requirements depending upon the specific end use, such as:

1. Military personnel's clothing, submariners, or other specialist forces
2. Fire entry and fire proximity suits
3. Sodium splash-resistant material for nuclear power stations
4. Low-risk fire-resistant coveralls for off-shore oil rigs
5. Clothing for aerospace industry

The role of fiber blending to enhance protective performance of clothing is addressed in relation to a few specific situations.

A. Military Personnel's Clothing

As an example, clothing for submariners where any fire could be disastrous, particularly from smoke and toxic fume emission, must be heat and fire resistant. Also, in submarines, washing facilities are limited; therefore, clothing must be able to absorb perspiration and odor without having drying problems. PBI fiber, because of its fire-resistant properties, good handle, and comfort (moisture absorbing) as well as odor-absorbing properties, is the answer. However, PBI fibers have low strength and abrasion resistance; hence, a blend of PBI with high-tenacity Kevlar would be more practical or a blend of oxidized acrylic (Panox, Celliox) with heat- and fire-resistant Kermel (aromatic polyamide hydrazide) and wool may be a useful combination—the wool providing resilience, comfort, and wear properties.

It is important to remember that the flammability behavior of a blend fabric cannot be predicted from the flammability characteristics of its component fibers. This is due to physical or chemical interaction of the thermal degradation products of the blend during heating or combustion. Some fiber blends exhibit synergism, for example, PBI–Cordelan (polychal fiber) or

PBI–wool have a higher LOI than the average of the component fibers, whereas some combinations (blends) are a disaster such as 90% Panox mixed with 10% FR polyester. Thus, blend specifications are based on actual flammability testing of blend fabrics.

Another example of military personnel's clothing is for special forces operating against hostage takers, where stringent criteria become operative. One important property requirement would be flexibility in flame conditions, because of the high mobility requirement of a surprise attack. For such clothing, using a high proportion of PBI or carbon precursor fibers such as Kynol or Panox blended with high-performance aramids (Kevlar) would perform satisfactorily as opposed to 100% fibers.

B. Industrial Protective Coveralls

Industrial protective coveralls may be composed of FR-cotton, FR-rayon, FR-wool, coated materials, or inherently heat- and fire-resistant fibers. PBI and aramid blends have exhibited excellent thermal, chemical, and textile properties for most critical industrial environments. Blends of 40/60 PBI/HS aramid, or 20/80 PBI/aramid are commercially available.

C. Fabrics for Flashover Fire Protection

Fire fighters' protective clothing is a complex multilayered ensemble and includes face covers (balaclava), gloves, helmets, and even self-contained breathing apparatus (SCBA).

The turnout coats are multilayered, with an impermeable FR outer shell, an intermediate vapor barrier, and an inner thermal-resistant liner or quilt. Nomex III was developed by DuPont as an inherently fire-resistant material for outer-shell fabric. DuPont has engineered a new blend of aramid fibers for high-risk protective clothing named Delta A. It is a blend of Nomex, Kevlar, and P140, a special cotton spun into a protective nylon filament. The clothing is claimed to combine non-break-open protection against flame and heat with protection against buildup of electrostatic charges.

Garments made from this new fabric are for use in high-risk environments, such as for structural fire fighting, the gas and chemical industries, and offshore platforms. The product could be valuable for rescue teams and military personnel handling ammunition and explosives, where the potential for creating a flashfire situation as the result of an electrostatic discharge is a real and serious problem. Garments made from the fiber have been undergoing tests in the North Sea oil fields (50).

D. Fabrics for Molten Metal Splash Resistance

Wool is the traditional protection against molten metal, although when Zirpro treated and chrome dyed, there is an increased propensity for the molten metal to adhere to the material. It does seem to follow, even in the case of wool, that making the material structurally more sound causes a slight deterioration in its metal-shedding properties, although its fire resistance is necessarily improved. The problem is not with the fabric's resistance to metal splash, but that the comfort factor is poor in the inevitably hot environment where the work can be arduous. Wool blends with aramids and pre-carbon fibers like Kynol and Panox and fabrics containing steel fibers have all been tried without success. Experiments indicate that the more resistant to heat a fiber is, the less likely it is to degrade; and if the fiber is stable, then the metal sticks. Within a very short period, the temperature on the inside of the fabric would reach 45°C and the worker would be in pain. Having tried most of the common synthetic fibers, the only two fibers that seem to perform reasonably well against the molten metal splash are wool and cotton. Of the two, wool would be preferred. However, special fabric structures such as double cloth, with a closely woven sheer twill made of FR-wool as a face fabric and a 100% FR-cotton back, are comfortable and molten metal splash resistant.

IV. CONCLUSIONS

Inherently heat- and fire-resistant fibers have been developed for protective clothing needs in such areas as fire fighting, gas, petrochemical and aerospace industries, and the military. Some commercially available and important fibers along with their applications have been discussed. Kynol fiber by Carborundum Company is based on a novolac precursor post-crosslinked with formaldehyde and cured to give a nonfusible, nonflammable fiber. It chars without melting at 550°C and carbonizes in a flame, evolving mostly CO_2 and H_2O. At carbonization, it retains 60% of its original weight with slight shrinkage. Nomex fiber also exhibits good flame resistance. It is difficult to ignite, generates little smoke, and is self-extinguishable when removed from a flame. However, Nomex may not provide complete flame protection in a catastrophe such as total immersion in a flame. Polybenzimidazole fiber is known to offer improved flame resistance under these conditions, with excellent comfort properties. Polyphenylene sulfide fibers exhibit a good combination of thermal, chemical, and textile properties and offer many industrial applications, particularly as filter media and replacement for glass fibers. On the other hand, glass fibers, especially the highly texturized Zetex fibers, are good contenders for asbestos replacement.

Fiber-blending technology in conjunction with unique woven and non-woven fabric structure has expanded the use of these fibers and enhanced protective clothing performance. Protective clothing for industrial safety and health is a dynamic area of research. No doubt more improvements will occur as the problems are better understood.

REFERENCES

1. A. M. Stoll and M. A. Chianta, *Aerospace Medicine 41*:1232 (1969).
2. B. V. Holcombe, *Fire Safety J. 6*:129 (1983).
3. J. H. Veghte, *Fire Service To-Day*, 16 (Aug. 1983).
4. B. V. Holcombe and B. N. Hoschke, Do test methods yield meaningful performance specifications?, *Performance of Protective Clothing*, ASTM STP 900 (R. L. Barker and G. C. Coletta, eds.), American Society for Testing and Materials, Philadelphia, 1986, pp. 327–339.
5. S. L. Kwolek, P. W. Morgan, and W. R. Sorenson, U. S. Patent 3,063,966 (1962).
6. S. L. Kwolek, U.S. Patent 3,671,542 (1972); U.S. Patent 3,819,587 (1974).
7. P. W. Morgan, *Macromolecules 10* (6):1381 (1977).
8. S. L. Kwolek, P. W. Morgan, J. P. Schaefgen, and L. W. Gulrich, *Macromolecules 10* (6):1390 (1977).
9. P. W. Morgan, *J. Polym. Sci., Polymer Symp. 72*:27 (1985).
10. H. Blades, U.S. Patent 3,767,756 (1973); U.S. Patent 3,869,429 (1975).
11. S. Ozawa, Y. Nakagawa, K. Matsuda, T. Nashihara, and H. Yunoki, U.S. Patent 4,075,172 (1978).
12. H. H. Yang, *Aromatic High Strength Fibers*, John Wiley, New York, 1989.
13. J. Preston, Aramid fibers, *Encyclopedia of Polymer Science and Engineering*, 3rd Ed., Vol. II, 1985, p. 383.
14. K. Hillermeier, Prospects of aramid as a substitute for asbestos, *Textile Res. J. 54*:575 (1984).
15. K. C. Brinker and I. M. Robinson, U.S. Patent 2,895,948 (1959).
16. H. A. Vogel and C. S. Marvel, *J. Polym. Sci. 50*:511 (1961).
17. H. Vogel and C. S. Marvel, U.S. Patent 3,174,974 (1965).
18. A. B. Conciatori, A. Buckley, and D. E. Stuetz, Polybenzimidazole fibers, *Handbook of Fiber Science and Technology*, Part A (Lewin and Preston, eds.), Marcel Dekker, New York, 1985, p. 3.
19. A. Buckley and G. A. Serad, Polybenzimidazoles, *Encyclopedia of Science and Technology*, Vol. II, 3rd Ed., McGraw-Hill, New York, (1985), pp. 572–586.
20. D. E. Stuetz, A. H. DiEdwardo, F. Zitomer, and B. P. Barnes, *J. Polym. Sci. Polym. Chem. Ed. 18*:987 (1980).
21. D. R. Coffin, G. A. Serad, H. L. Hicks, and R. T. Montgomery, *Textile Res. J. 52*:486 (1982).

22. R. N. DeMartino, Comfort properties of polybenzimidazole fiber, *Textile Res. J. 54*:517 (1984).
23. R. M. Stanton, Heat Transfer and Flammability of Fibrous Materials (AFML-TR-70-238), Air Force Materials Laboratory, Wright-Patterson Air Force Base, Ohio (1971).
24. R. M. Stanton, *The protective characteristics of PBI and Nomex coveralls in JP-4 fuel fires* (AFML-TR-73-27), Air Force Materials Laboratory, Wright-Patterson Air Force Base, Ohio, 1973, 11, 12, 13, 23, 41, 42.
25. R. M. Stanton, S. Schulman, and J. H. Ross, *Evaluation of PBI and Nomex II for Air Force flight suites* (AFML-TR-7328), Air Force Materials Laboratory, Wright-Patterson Air Force Base, Ohio, 1973, pp. 5, 6, 32, 63, 72, 73, 84, 105–108.
26. R. H. Jackson, PBI fiber and fabric—Properties and performance, *Textile Res. J. 48*:314 (1978).
27. R. E. Bouchillon, Protective performance of polybenzimidazole-blend fabrics, *Performance of Protective Clothing*, ASTM STP 900 (R. L. Barker and G. C. Coletta, eds.), American Society for Testing and Materials, Philadelphia, 1986; pp. 389–404.
28. ASTM, *Standard Test for Evaluating Heat Transfer Through Materials Upon Impact of Molten Substances*, Draft No. 2, ASTM F23-20 on Physical Properties, Subcommittee of ASTM Committee F-23 on Protective Clothing, American Society for Testing and Materials, Philadelphia, Oct. 8, 1982.
29. L. Benisek and G. K. Edmondson, Protective clothing fabrics, Part I, Against molten metal hazards, *Textile Res. J. 51*:182 (1981).
30. L. Benisek, G. K. Edmondson, and W. A. Phillips, Protective clothing—evaluation of wool and other fabrics, *Textile Res. J. 49*:212 (1979).
31. J. S. Hayes, Jr., *Encyclopedia of Chemical Technology* (Kirk-Othmer), 3rd Ed., Vol. 16, John Wiley & Sons, New York, 1981, p. 125.
32. J. Economy and L. Wohrer, Phenolic fibers, *Encyclopedia of Polymer Science and Technology*, Vol. 15 (N. M. Bikales, ed.), Wiley-Interscience, New York, 1971, pp. 370–373.
33. J. Economy, L. C. Wohrer, F. J. Frechette, and G. Y. Lei, *Appl. Polym. Symp. 21*:81 (1973).
34. J. Economy, Phenolic fibers, *Flame-Retardant Polymer Materials*, Vol. 2 (M. Lewin, S. M. Atlas, and E. M. Pearce, eds), Plenum Press, New York, 1978, pp. 210–219.
35. J. S. Hayes, Jr., Novoloid Nonwovens, 1985 Nonwoven Symposium, TAPPI, Atlanta, 1985, pp. 257–263.
36. J. H. Ross, *Fire and Materials* (London) *4* (3):144 (1980); Abstracted in Textile Flammability Digest 2/81:3 (1981).
37. *Federal Register* 49 (209):43188 (October 26, 1984).
38. H. W. Hill, Jr., Polyphenylene sulfide: Stability and long-term behavior, *Durability of Macromolecular Materials, ACS Symposium Series 95*, ACS, Washington, D.C., 1979, pp. 183–197.

39. J. Scruggs and J. Reed, Polyphenylene sulfide fibers, *Handbook of Fiber Science and Technology, Part A, High Technology Fibers*, Vol. III (M. Lewin, J. Preston, eds.), Marcel Dekker, New York, 1985, pp. 335–348.
40. H. W. Hill and J. T. Edmonds, Jr., Properties of polyphenylene sulfide coatings, *Advances in Chemistry Series, Polymerization Reactions and New Polymers* (N. A. Platzer, ed.), ACS, Washington, D.C., 1973, pp. 80–91.
41. R. W. Lenz and W. K. Carrington, Phenylene sulfide polymers, I. Mechanism of the Macallum polymerization, *J. Polym. Sci. 41*:333 (1959).
42. G. C. Bailey and H. W. Hill, Polyphenylene sulfide: A new industrial resin, *ACS Symposium Series 4*, ACS, Washington, D.C., 1972, pp. 83–99.
43. D. G. Brady and H. W. Hill, Jr., How some high-performance resins stack up in chemical resistance, *Modern Plastics 51*:60 (1974).
44. H. L. Needles, Mineral and metallic fibers, *Textile Fibers, Dyes and Finishes*, Noyes Publications, N.J. 1986, pp. 116–119.
45. F. A. Mennerich, New textile uses for glass fiber yarns, *Modern Textiles 43–45*:67 (December 1962).
46. F. S. Galasso, Glass fibers, *Advanced Fibers and Composites*, Gordon and Breach Science Publishers, New York, 1989, pp. 98–109.
47. M. E. Carter, Chapter 10: Glass, *Essential Fiber Chemistry*, Marcel Dekker, New York, 1971, pp. 183–199.
48. R. F. Caroselli, Glass textile fibers, *Man-Made Fibers, Science and Technology* (H. F. Mark, ed.), Interscience Publishers, New York, 1968, pp. 425–454.
49. W. Smith, Akzo introduces two high performance yarns, *Textile World 141*:39 (1991).
50. *Fire and Flammability Bulletin 9* (March 1991).

9

Protective Clothing Effective Against Biohazards

TYRONE L. VIGO Agricultural Research Service, U.S. Department of Agriculture, New Orleans, Louisiana

I. INTRODUCTION

The recognition that clothing and garments may serve as a carrier of diseases dates back to the Middle Ages when clothing was burnt to prevent the spread of the dreaded bubonic plague. However, the first scientific demonstration of the relationship between fibrous materials and disease was made by the famous British surgeon Joseph Lister. He impregnated bandages with carbolic acid (phenol, the first antiseptic) to prevent infection in wounds [1]. Since that time, fiber science and the allied fields of microbiology, physical and organic chemistry, and polymer science have advanced and led to the production of textiles that serve as biobarriers. Most of the protective clothing is currently used in hospitals and other confined environments conducive to cross-infection or transmission of diseases caused by microorganisms. Moreover, the recent epidemic of devastating and lethal virus-related diseases [such as hepatitis and human immunovirus (HIV)] have led to more stringent guidelines for protection of personnel with appropriate clothing and disposable materials in such environments. The discussion focuses on current knowledge of the factors that cause adherence and survival of microbes on clothing and fibrous materials, strategies to provide effective protective clothing against biological hazards, and the major end uses for such protective clothing.

II. MICROBIOLOGY OF THE SKIN–CLOTHING INTERFACE

The microbiology of the skin–clothing interface is important in determining the propensity to biohazards from causative pathogenic microorganisms. Important factors and their scope are summarized in Table 1 and include adhesion to substrates or to fibrous materials, persistence of microorganisms on the skin and on clothing, and the microbial ecology of the skin–clothing interface.

Table 1 Microbiology of the Skin–Clothing Interface—Important Factors and Attributes

Subtopic of importance	Relevant features or parameters
Adhesion to fibrous substrates	Microbes approach surfaces by modes of transport or active movement Initial adhesion occurs at some point Attachment of microbes achieved by fibril or polymer formation between microbe and fiber surface Microcolonies or biofilms adhere to fibrous surface
Persistence of microorganisms	Synthetic fibers retain more odor-causing bacteria and dermatophytic fungi than do natural fibers Aerosol contamination at low RH most conducive for microbial survival Inactivation of surfaces contaminated with hepatitis virus difficult
Skin microflora relationship to type of microbe	Microbe type and amount varies markedly from one part of the body to another Odor-causing bacteria prevalent on axillae, legs, and arms Groin and feet have more pathogenic bacteria, yeasts, and dermatophytic fungi than other parts of the body Rapid microbial growth conducive in body area that are wet or dark and/or contain organic matter that serve as microbial nutrients

A. Adhesion of Substrates

The mechanisms and factors that promote attachment of microflora to fibrous surfaces and other similar substrates are of paramount importance in protection of personnel against biohazards and infectious diseases. If surfaces can be designed to minimize or alleviate adhesion of microbes, then the materials will be more effective in many situations and environments. Surprisingly little information is available on this topic. However, there are related studies on how bacteria and other microorganisms adhere to nonfibrous substrates.

Mechanisms and physiological significance of bacterial adhesion are discussed in a comprehensive monograph [2]. Even though there is no specific discussion of how bacteria and other deleterious microorganisms become attached to fibrous surfaces, there is some discussion of how such attachment takes place on inert plastic prostheses placed in the human body (e.g., catheters and cardiac pacemakers) [3] and on the properties of nonbiological surfaces that are important for such adhesion [4]. Bacteria and other microorganisms adhere to all types of surfaces by a sequence of four processes shown in Figure 1 [5]. Bacteria may approach surfaces by either diffusive transport, convective transport, or active movement. The initial

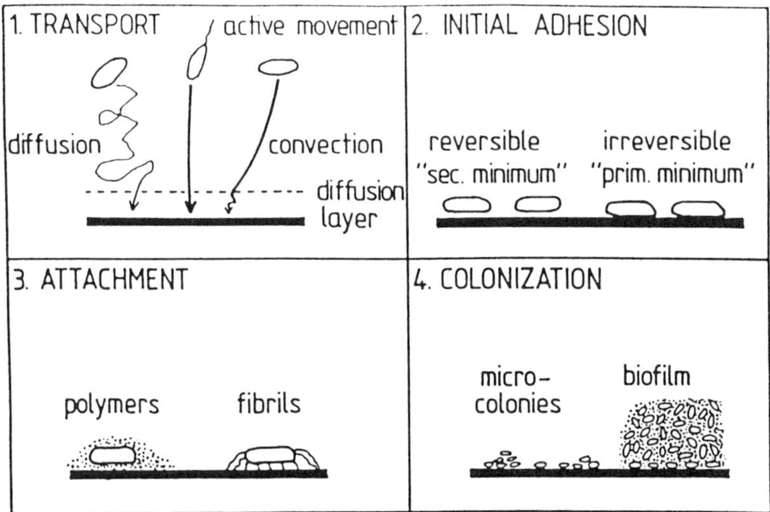

Figure 1 Schematic representation of the sequencing steps in the colonization of surfaces by microorganisms. (From Ref. 5.)

steps of bacterial adhesion may be described by colloid chemical theories such as the DLVO theory (Derjaguin–Landau–Verwey–Overbeek theory of colloidal stability). Such theories describe changes in Gibbs energy as a function of the distances between two bodies (in this instance the microorganisms and the surface); the overall interaction is the sum of Van der Waals and electronic interactions. Ionic strength and distances of separation of the two bodies are the most important factors in predicting initial adhesion [5]. Such initial adhesion may be reversible in that when microorganisms exhibit Brownian motion, they can be readily removed by mild shear forces or adhesion may be irreversible when no Brownian motion occurs. The actual attachment step is achieved by the bacterium forming either fibrils or polymers to form strong links between the cell surface and the solid surface. The bacteria then form microcolonies or biofilms that adhere to the solid surface forming a biofilm comprised of naturally occurring polysaccharides and related structures. Although there may be differences in the chemical nature of polymers from one bacterial species to another, biofilm formation is a general phenomenon whereby the microbes affix themselves to a surface. Adherence of the pathogenic bacterium *Staphylococcus aureus* to a cardiac pacemaker is shown in Figure 2.

When the bacteria form an embedded matrix with the substrate via a biofilm, this gives the bacteria the capability of complexing or rendering ineffective antibiotics and/or antimicrobial chelating agents. The only effective methods for removing biofilms from surfaces after they are formed are (1) the use of oxidizing agents (such as hypochlorite) and (2) physical removal at low temperatures via ice formation to destroy the physical integrity of the biofilm. Oxidizing agents are effective not only because they kill the bacteria but also because the biofilm is depolymerized and thus becomes physically detached from the surface.

A fundamental study on the surface thermodynamics of bacterial adhesion of five different bacterial strains with five different polymeric surfaces was conducted [6]. It was determined that the extent of bacterial adhesion was determined by the surface properties of all three phases: surface tensions of the adhering particles, of the suspending liquid medium, and of the polymeric substrate. Adhesion to hydrophobic surfaces was more extensive when the surface tension of the bacteria was larger than that of the liquid medium; this trend was reversed when the surface tension of the liquid medium was larger than that of the bacteria. A later study demonstrated that the degree of adhesion of 16 different bacterial strains on a solid sulfated polystyrene surface was proportional to their contact angle with aqueous solutions of 0.1 *M* NaCl (i.e., the higher the contact angle of the bacterium with the salt

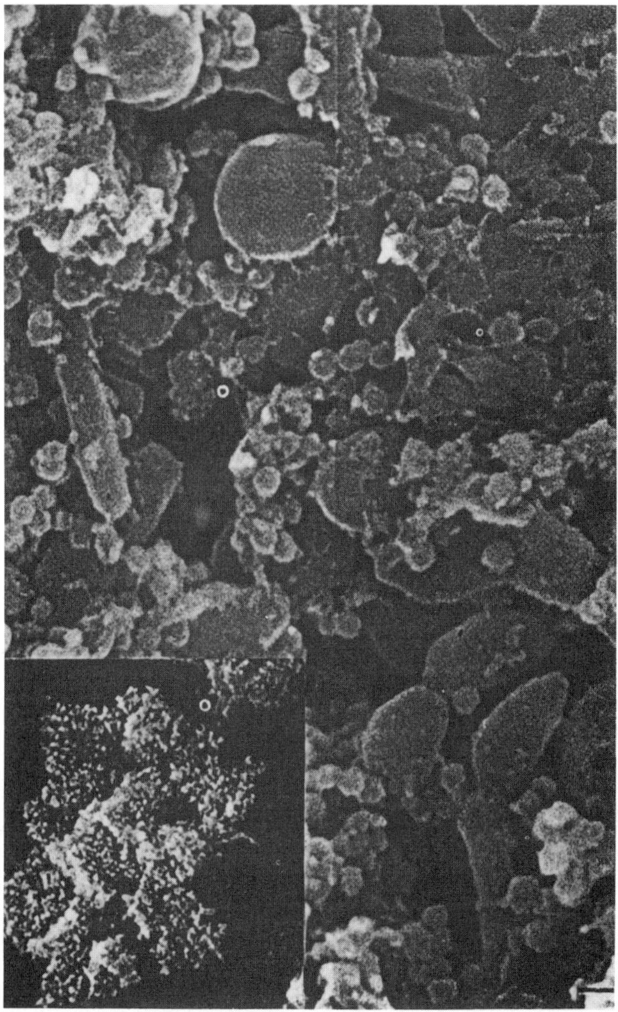

Figure 2 SEM of microorganisms within a huge accretion (inset) that had developed on the plastic surface of an intracardial Hickman catheter. This accretion was composed of coccoid bacteria and yeastlike cells. Laboratory cultures yielded *Staphylococcus epidermidis* and *Candida albicans*, and the patient developed a bacteremia from which these two organisms were also cultured. (Adapted from Ref. 3.)

solution, the greater was its adhesion to polymer surfaces [7]. Adhesion of the yeastlike fungi *Candida albicans* to plastic surfaces appears to be promoted by the same surface phenomena that occur with adhesion of bacteria to such surfaces [8].

The adhesion of a representative gram-positive bacterium (*S. aureus*) to six films (cellulose acetate, nylon 66, polyester, polyethylene, polypropylene, and Teflon) and four fabrics (cotton, nylon, polyester, and polypropylene) was measured after an initial hour contact of these surfaces with aqueous suspensions of the bacteria [9]. Although there was a good correlation between bacterial adhesion and increasing hydrophobicity of the film surfaces, no clear correlation was observed between the surface wettability of fabrics and bacterial adhesion. Thus, further studies are warranted to elucidate and characterize factors of fibrous surfaces that promote or minimize adhesion by bacteria and other microorganisms.

Attachment of viruses to solid surfaces has been critically reviewed and occurs by adsorption rather than adhesion [10]. This mode of attachment occurs due to the small size of a representative virus relative to bacteria and other microorganisms. However, the DLVO theory advanced for initial adhesion of bacteria is also applicable to adsorption of viruses, because the latter exhibit colloidal behavior. The two most important factors that influence virus adsorption to a surface are isoelectric point and hydrophobicity; because viruses and surfaces differ widely in these parameters, specific studies are warranted to determine affinity of viruses for various fibrous and related polymeric surfaces.

B. Persistence of Microorganisms

Most of the comprehensive studies on the persistence time of pathogenic bacteria, fungi, and viruses on fibrous materials were conducted about two decades ago and have been critically reviewed [11]. Fiber type has some influence on the persistence of microorganisms on textiles. Synthetic fibers generally retain more odor-causing bacteria (such as *Staphylococcus epidermidis*) and dermatophytic fungi (such as *Trichophyton interdigitale)* than do natural fibers. However, persistence time of pathogenic bacteria and viruses on fibers are more dependent on relative humidity (RH) and the mode by which the fabrics become contaminated. Aerosol contamination by pathogenic *S. aureus* and polio virus of fabrics subsequently held at low humidity (35%) was the most conducive for survival of the microorganisms, whereas contamination by microorganisms of fabrics in solution subsequently held at high humidity (78%) led to the shortest persistence times [12,13]. Even though there are no specific studies on the persistence of the HIV and hepatitis viruses on fibers, generally it has been observed that inactivation of hepatitis viruses on contaminated surfaces is much more difficult than corresponding inactivation of surfaces contaminated by the HIV virus.

C. Relationship of Skin Microflora to Microbe Type

The microbial ecology of the skin–clothing interface is fairly well documented and has been critically reviewed [11]. Healthy, unbroken skin serves as a mechanical barrier to infection and always has a population of microorganisms that differ in amount and type from one part of the body to another, with different age groups, and with different geographic locations [14]. For example, areas such as the groin and feet contained greater amounts of *S. aureus*, gram-negative bacteria, yeast, and dermatophytic fungi than other body areas [15]. The axillae, legs, and arms contained the highest concentrations of odor-causing bacteria such as *S. epidermidis* and coryneforms.

Rapid microbial growth on the skin and skin–clothing interface usually occurs in wet and dark areas, where the skin is broken and/or parts of the body are contaminated with body effluents and other organic materials that serve as microbial nutrients. Cross-infection or transmission of pathogenic microorganisms is much more favorable in confined environments such as hospitals, nursing homes, day care centers, and similar establishments. For example, clothing in the inguinal and perineal areas soiled by urine and feces promotes the growth of bacteria such as *E. coli* and *P. mirabilis* and thus increases the chance for diaper rash and associated infections [16]. The recent emergence of resistant strains of *Mycobacterium tuberculosis* and the prevalence of insidious viruses such as hepatitis has led to the development of guidelines for appropriate protective clothing for health care workers and medical personnel (discussed later in this chapter).

III. APPROACHES FOR PRODUCING ANTIMICROBIAL FIBERS

For modern-day protective clothing, three general strategies (Table 2) that afford the best protection against biohazards and deleterious microorganisms have evolved: (1) two physiocochemical techniques—formation of a physical biobarrier against microbial infiltration or transmission by use of appropriate fabric constructions and use of impermeable or microporous coatings and microencapsulation of active agents in fibrous assemblies—(2) one chemical approach—incorporation of active antimicrobial agents in spinning baths

Table 2 General Strategies for Protecting Textiles and Their Users Against Biohazards

Technical approach	Relevant features or parameters
Physicochemical techniques	Physical barriers impermeable to both liquid and vapor and/or microporous coatings permeable only to vapor but not to liquids
	Microencapsulation of bioactive agents that operate by controlled release
Chemical methods	Controlled release of active biocide or biostat applied by finishing methods
	Subclasses including chelating agents, oxidizing and/or reducing agents, and various types of antibiotics
Multifunctional property fabrics	Flame-retardant/antibacterial fabrics
	Fabrics with multifunctional properties including antimicrobial resistance
	Clothing resistant both to chemical and biohazards

when fibers are produced and topical application of these agents onto fabrics. Depending on the application, one or more of these strategies may be required to produce protective clothing effective against biohazards.

A. Physicochemical Methods

Formation of physical barriers to prevent microbial contamination and/or transmission is normally achieved by choice of fabric construction, use of impermeable or microporous coatings, and/or waterproofing garments. Fabrics that generate little or no particles or lint were produced initially for prevention of dust particles in computer rooms and a few biomedical applications. However, recently it has been observed that it is important to also have such garments because many infectious diseases may be transmitted by airborne microorganisms [17]. Earlier garments for clean room use were normally composed of hydrophobic fibers such as polyethylene or polypropylene and lacked important comfort features. This has led to the development of more complex assemblies that contained microporous coatings in combination with fabrics exhibiting little particle generation [18].

Similar strategies have evolved for producing surgical gowns, drapes, and related materials that are resistant to penetration by biological fluids (such as blood) to prevent transmission of infectious diseases. Garments and materials initially used for these purposes were waterproofed and effective against "bacterial strikethrough" but lacked important aspects of thermal comfort. These requirements have been critically reviewed [19], and the various classes of water-repellent agents listed (waxes and fluorocarbons). Again, microporous, breathable coatings (such as Gore-Tex) are being considered as viable alternatives to waterproofed fabrics, because they allow transmission of water vapor but do not allow entry of liquids due to their pore structure [20]. The emergence of hepatitis and HIV infections in the last 10 years has led to the recent ruling by OSHA [21] that health care and other personnel wear appropriate protective clothing when handling body fluids and/or contaminated laundry and other fomites that may transmit these insidious viruses. Thus, use of such clothing protective against biohazards is now mandated in many situations and should dramatically increase the research activity to produce garments suitable to meet these new regulations.

The first use of microencapsulation to incorporate an antibacterial agent in fibrous substrates was disclosed almost 20 years ago; a quaternary ammonium salt was incorporated into mattresses and was proven to be effective against the representative gram-positive bacteria *S. aureus* even after 5 years of use [22]. This method was based on a reservoir of antimicrobial agent in one layer bonded to another layer of material by an adhesive. More effective methods of microencapsulation to release active biocidal agents have now been devised. A representative example is the microporous acrylic fiber Actipore that contains agents that can selectively diffuse through pores of different sizes, thus controlling the rate of the release of the active agent for specific applications [23]. Although microencapsulation is still relatively expensive compared to other methods of imparting antimicrobial activity, it offers the most versatility in choice of agents effective against specific microorganisms and has the capability of controlling the desired rate of release of the antimicrobial agent.

B. Chemical Methods

The earliest methods of imparting antimicrobial activity to fibers used simple deposition of the active agent on the substrate, usually from an organic solvent. These approaches have been discussed in a historical review on antimicrobial polymers and fibers [24]. Since 1960, it has been recognized that the most effective method of imparting activity to fibers against pathogenic microorganisms is by some form of controlled release. In a landmark paper by Gagliardi [25], different chemical approaches (covalent bond formation, chelation, metastable bonds, and the presence of a reservoir of active antibacterial agent) are discussed. Since that time, most active agents that are durable to laundering or other forms of refurbishing have been incorporated into fibers either during the spinning bath (macroencapsulation) or as topical finishes as part of a graft, homo-, co-, or crosslinked polymer [24].

The commercial production of the broad spectrum antimicrobial poly (vinyl alcohol) fiber was the first successful demonstration of the concept of controlled release; the active agent was 5-nitrofurylacrolein that was hydrolyzed from the acetal of the alcohol groups in the fiber [26]. Cellulose and cellulose blend fabrics having antibacterial activity durable to repeated laundering were produced by the slow release of a zinc peroxide polymer incorporated into the fabric [27]. Another significant development was the attachment of an inherently biostatic coating (based on a polymeric silicone containing a pendant quaternary ammonium group) to all types of fibrous surfaces; this is the only commercial product that is not based on the controlled release concept for its antimicrobial activity [28]. The most recent approach for producing antibacterial fibers that has received attention is the incorporation into fibers of synthetic zeolites treated with antibacterial transition metal ions (e.g., silver, mercury, and zinc) [29]. Although this approach has limitations due to its expense and to the incorporation of such agents only in spinning baths, these zeolite-modified fibers appear to be effective against several fungi and bacteria. Another recent approach that has commercial appeal is the incorporation of a broad-spectrum antimicrobial phosphate into the poly (vinyl chloride) backing of carpets that provides activity lasting several years due to slow rate of release of the active agent [30]. Because carpets are used in hospitals and other confined environments conducive to transmission of disease-producing microorganisms, evaluation of this approach for public health and health care applications is desirable and of benefit.

C. Multifunctional Properties and Uses

Some materials have been produced that have other improved functional properties in addition to antimicrobial protection or that are useful against more than one mode of biological contamination. Subsequent examples that are discussed represent a diversity of approaches and properties.

Flame-retardant cotton fabrics containing a phosphonium salt (THPC) were also observed to have excellent antibacterial activity even after 50–100 launderings against a diverse group of pathogenic bacteria (e.g., *Salmonella typhosa* and *Shigella dysenteriae*). Unfortunately, no further studies were conducted on these dual-property fabrics to determine the active antibacterial agent or the minimum amounts of the flame-retardant finish required to impart such antibacterial activity [31]. Kanebo has developed an antistatic polyester-based fabric (containing electrically conductive fibers) that is also acceptable for clean room use [32].

Incorporation of crosslinked polyethylene glycols into all types of fibers and fabric constructions produces materials with several improved functional properties useful in biomedical and clean room environments: thermal comfort, reduced static charge, no lint loss or particle generation, and antimicrobial activity [33,34]. The Canadian Ministry of Defense has recently patented protective clothing that is resistant to both chemical and biological hazards; it is composed of several layers of laminated material that render it impermeable to liquids and particulate matter while being vapor permeable to provide thermal comfort during wear [35]. It is highly probable that other innovative approaches for producing clothing with multifunctional properties will be discovered and marketed.

IV. EFFECTIVE DECONTAMINATION, DISINFECTION, AND STERILIZATION

Protective clothing may need to be sterilized, disinfected, sanitized, and/or decontaminated from microorganisms. The extent of biological cleanliness will depend to a large extent on its end use and whether it is a disposable item. Table 3 lists historical and current definitions of the terms *disinfection, sanitizer*, and *sterilization*. Decontamination usually denotes total or substantial removal of infectious or disease-causing microorganisms from materials to a level considered safe from a health or medical perspective.

Sterilization means that a surface or material is ''germ-free'' prior to its use. Although sterility is normally associated only with bandages and materials that will be placed inside the body, there may be instances where protective clothing is required to be sterile. Some examples would be sterile hospital gowns and scrub suits for patients and surgeons, respectively, to minimize infection by or from either party. More commonly, protective clothing effective against biohazards should have the capability of being decontaminated by appropriate laundering and/or use of disinfectants and if it will be worn again or if this is part of the procedure for its proper disposal. Sanitized surfaces are only minimally protective in such environments, and this level of decontamination is probably not suitable in environments where transmission of infectious and potentially lethal diseases may occur.

Table 3 Selected Historical and Current Definitions of Antimicrobial Action

Term	Original definition (year or time)	Current definition
Disinfectant	Any agent that would free anything from infection (seventeenth century)	Products that kill the growing forms, but not necessarily the resistant spore forms, of microorganisms on inanimate objects by decontamination
Sanitizer	Disinfectant without any residue harmful to later users of the article or product that eliminates any aesthetically objectionable contaminants (1945)	Agent commonly used with that are applied to inanimate objects, which reduces number of bacterial contaminants to safe levels, as judged by public health laws
Sterilization	Act or process or freeing from all living microorganisms (1932)	Same as original definition

Adapted from Ref. 24.

A. Physical Methods

The concept of sterilization dates back to the turn of the century when it was discovered that either heat or shortwave length radiation (such as x-rays) were suitable for sterilizing fibers and other surfaces. Later, it was determined that ultraviolet light was also suitable as a method of sterilization. Commercial methods currently employed for producing sterile bandages and other fibrous materials use ionizing radiation or dry or moist heat.

B. Chemical Methods

Chemical methods of sterilizing surfaces are employed when the material (in this case a garment or fibrous substrate) is degraded by heat or shortwave length radiation. Since chemical sterilization normally occurs at ambient rather than elevated temperatures, it is commonly referred to as *cold sterilization*. Gaseous sterilization with ethylene oxide or formaldehyde was first used in the 1920s [24] but has fallen into some disfavor because of the potential carcinogenic effects of ethylene oxide and mutagenic effects of formaldehyde.

For disinfection and decontamination of clothing, there are numerous chemical agents that have been proven to be effective and that are still widely used. The most important consideration is whether the organism is a pathogenic bacteria or a fungus or a bloodborne pathogen. The latter group comprises debilitating or lethal viruses such as hepatitis, HIV, syphilis, and malaria. Removal of disease-causing bacteria from clothing is readily achieved in most instances by hot laundering with added disinfectants (such as bleach or hypochlorite) followed by tumble-drying. A variety of other disinfectants suitable for removing bacteria and related microorganisms from garments have been critically reviewed and include hydrogen peroxide, peracids, perborate, quaternary ammonium salts, glutaraldehyde, halogenated guanidinium salts, and metal salts of 2-mercaptopyridine-1-oxide [11].

Detailed laundering studies on the survival of selected viruses (primarily polio and vaccinia) on fabrics were conducted before the discovery of the HIVs and the higher incidence of infection by various forms of hepatitis. It was determined that hot water wash with detergents markedly reduced the amount of detectable polio virus on a variety of fabrics. However, it was also observed that viable polio and vaccinia viruses remaining on dried fabrics were readily transferred to uncontaminated fabrics by dry, random tumbling contact [36,37]. The recent OSHA standard on protection of personnel from bloodborne pathogens (primarily various forms of hepatitis and HIVs) includes specific guidelines for laundering of contaminated clothing and items. The standard also addresses the possibility of transmission from infected garments to noninfected garments and environments [21]. This standard recommends minimum handling of such laundry, collection of such items in special leak-proof bags, and proper control of laundering of these items in specially designated facilities either on- or off-site. If the garments, protective clothing, towels, or other textile materials are not suitable for reuse, guidelines also are given for proper disposal of these items.

In a recent study, it was observed that only 3 of 20 common disinfectants (hypochlorite or bleach, glutaraldehyde, and a quaternary ammonium salt composition containing substantial amounts of hydrochloric acid) were effective in killing hepatitis A viruses on environmental surfaces [38]. Although interpretation of these results may not be totally applicable to infected garments, it certainly indicates the need for detailed and current studies of disinfectants and laundering conditions suitable for completely killing such insidious and lethal viruses, both on the fabrics and in the waste-water effluents.

V. CONSIDERATIONS FOR SPECIFIC END USES

Protective clothing against biohazards may be categorized according to the three basic types of biohazards and the occupation and/or environment in which it is used. Table 4 lists this information; much of this data was taken from the OSHA standard that lists relevant occupations at risk to bloodborne pathogens [21].

A. Airborne Transport of Particles and/or Microbes

Protective clothing (such as gowns, masks, and gloves) to minimize release of particles was initially designed for clean rooms to prevent contamination of computer chips and other products that required dust-free environments. This type of clothing was later used in operating rooms to prevent or minimize granuloma infections of patients from lint or fiber particles in the environment. Although there are no specific guidelines for minimizing transmission

Table 4 Protective Clothing Effective Against Biohazards—Type and Situation

Type of biohazard	Relevant occupant or environment
Airborne transfer of skin, microbes, and/or fiber particles	Operating room personnel and patients Computer clean rooms Aerospace and defense
Bloodborne pathogens, particularly hepatitis viruses and HIV	Health care workers (includes various professional and support personnel) Industrial facilities, research laboratories, and schools
Other pathogenic microorganisms present in body fluids	Law enforcement, fire and rescue, and correctional facilities Funeral homes Linen services and commercial laundries Waste disposal
Pathogenic bacteria, fungi, and other microorganisms	Health care workers, personnel in confined environments where infectious diseases are present Carpets in hospitals and related environments Aerospace and defense

of airborne pathogens, a secondary purpose of protective clothing that does not readily form lint or generate particles is the ability to minimize transmission of pathogenic microorganisms from the skin or as part of lint particles generated in confined environments. In the recent OSHA standard [21], there appears to be no incidence of insidious bloodborne pathogens transmitted in air. Nevertheless, the use of masks and lint-free protective clothing may be appropriate in biomedical and military situations when pathogenic bacteria and other disease-causing microorganisms are known to be transmitted by aerosol modes. Protective clothing for cleanup of biological spills of microorganisms known to be transmitted by air is also appropriate.

B. Liquid Transport of Pathogens

The emphasis on liquid-proof or liquid-resistant protective clothing and related textile items is that it be resistant to the hepatitis and HIV viruses described in detail in the recent OSHA standard [21]. Relevant protective clothing and related items are: gowns, gloves, masks, respirators, and goggles. U.S. population estimated to be at risk and require some form of protective clothing when exposed to bloodborne pathogens is 5.6 million for the HIVs and 2.5 million to 3.0 million for the hepatitis viruses; the latter figure is lower because prior exposure to hepatitis or vaccination may result in immunity to the hepatitis viruses. The majority of personnel at risk are health care workers that are located at the following facilities: hospitals, dental offices, physicians' offices, medical and dental labs, nursing homes, residential care facilities, home health and hospice care, hemodialysis and drug treatment centers, public clinics and blood banks, and industrial and correctional facilities. Other occupations and facilities (such as law enforcement and rescue) are listed in Table 4.

The prime requirement to avoid penetration of liquids to the skin is that the blood or other contaminated body fluids not strike through the protective garment. Thus, most gowns and other outer protective garments should be liquid-proof but also allow the transmission of vapor to provide some measure of thermal comfort. As noted earlier, there are some garments that are composed of microporous coatings that provide comfort yet afford resistance to body fluids. In the case of gloves, any feature that would minimize the penetration of needles through the skin is considered highly desirable. Even though there is no material that is inherently resistant to penetration by needles, the use of cut-resistant glove liners made of polyolefin fibers affords much better protection than disposable gloves or other gloves that may be readily cut or torn [39].

C. Antimicrobial Proofing of Protective Clothing

There may also be environments and situations where it is appropriate to have protective clothing that contains an active antimicrobial agent. Treatment of carpet and other textile flooring material in hospital and other confined environments is also most appropriate, since it has been demonstrated that carpets with an antimicrobial finish are highly effective in controlling growth and transmission of pathogenic bacteria in hospitals and other confined environments [40].

Two primary considerations for application of an antimicrobial agent to protective clothing or other textile items is whether or not the fibrous material is disposable or refurbishable (cleaned by shampooing or laundering) and which microorganisms may be problematic for the user or wearer of the protective clothing.

If the item or garment is disposable, then the antimicrobial agent need not be durable to prolonged laundering or cleaning. However, if durability to commercial hospital laundering or cleanability is a requirement, the number of modified textiles that meet these requirements are relatively small. Very few antimicrobial textiles that have prolonged durability to laundering or other methods of cleaning exist. If the types of microorganisms to which one is exposed are known with some certainty, then the modified protective clothing should be effective against that microorganism under representative end-use conditions. Conversely, if there are several types of pathogenic microorganisms, then the modified protective clothing should contain a broad-spectrum antimicrobial agent effective against many types of bacteria, fungi, and viruses.

VI. CONCLUSIONS

There are three major modes of producing clothing and other protective gear effective against biohazards: (1) selection of fabric construction, fiber type, application of coatings, and other techniques to prevent generation of fiber or other particles; (2) formation of a solid barrier to liquids to prevent contamination by pathogenic body fluids; and (3) imparting antimicrobial activity to fibrous substrates by a variety of physicochemical and/or chemical techniques. The most important consideration is to match the biohazard effectiveness of the modified garment to the types of microorganisms encountered in a particular environment or situation. It is probable that more suitable garments will be designed to address the complex needs of protection of many personnel from various biohazards. Further studies are needed to determine the factors that minimize adhesion and persistence of deleterious microorganisms in protective clothing, or that maximize removal of such organisms from contaminated clothing. To accomplish this objective, more fundamental investigations on the interaction of microbes with fibrous surfaces should be conducted.

REFERENCES

1. J. Lister, On the antiseptic principle in the practice of surgery, *The Lancet* 2:353,668 (1867).
2. D. C. Savage and M. Fletcher, eds., *Bacterial Adhesion: Mechanisms and Physiological Significance*, Plenum Press, New York, 1985.
3. J. W. Costerton, T. J. Marrie, and K.-J. Cheng, *Bacterial Adhesion: Mechanisms and Physiological Significance* (D. C. Savage and M. Fletcher, eds.), Plenum Press, New York, 1985, Chapter 1.
4. G. Loeb, *Bacterial Adhesion: Mechanisms and Physiological Significance* (D. C. Savage and M. Fletcher, eds.), Plenum Press, New York, 1985, Chapter 5.
5. M. C. M. vanLoosdrect, J. Lyklema, W. Norde, and A. J. B. Zehnder, Influences of interfaces on microbial activity, *Microbiol. Rev. 54* (1):75 (1990).
6. D. R. Absolom, F. V. Lamberti, Z. Policova, W. Zingg, C. J. van Oss, and W. Neumann, Surface thermodynamics of bacterial adhesion, *Appl. Envir. Microbiol. 46* (1):90 (1983).
7. M. C. M. vanLoosdrecht, J. Lyklema, W. Norde, G. Schraa, and A. J. B. Zehnder, The role of bacterial cell wall hydrophobicity in adhesion *Appl. Envir. Microbiol. 53* (8):1893 (1987).
8. R. A. Calderone and P. C. Braun, Adherence and receptor relationships of *Candida albicans, Microbiol. Rev. 55* (1):1 (1991).
9. Y.-L. Hsieh and D. Timm, Relationship of substratum wettability measurements and initial *Staphylococcus aureus* adhesion to films and fabrics, *J. Colloid Interface Sci. 123* (1):275 (1988).
10. C. P. Gerba, Applied and theoretical aspects of virus adsorption to surfaces, *Adv. Appl. Microbiol. 30* (A. I. Laskin, ed.), Academic Press, New York, 1984, p. 133.
11. T. L. Vigo and M. A. Benjaminson, Antibacterial fiber treatments and disinfection, *Textile Res. J. 51* (7):454 (1981).
12. L. J. Wilkhoff, L. Westbrook, and G. J. Dixon, Factors affecting the persistence of *Staphylococcus aureus* on fabrics, *Appl. Microbiol. 17* (2):268 (1969).
13. G. J. Dixon, R. W. Sidwell, and E. McNeil, Quantitative studies on fabrics as disseminators of viruses. II. Persistence of poliomyelitis virus on cotton and wool fabrics, *Appl. Microbiol. 14* (2):183 (1966).
14. W. E. Kloos and M. S. Musselwhite. Distribution and persistence of *Staphylococcus* and *Micrococcus* species and other aerobic bacteria on human skin, *Appl. Microbiol. 30*:381 (1975).

15. M. J. Marples, *The Ecology of the Human Skin*, Charles C. Thomas Publishers, Springfield, Mo., 1965.
16. P. J. Radford, Application and evaluation of antimicrobial finishes, *Amer. Dyest. Reptr. 62* (11):48 (1973).
17. M. J. Dyer, Fabrics for clean rooms, *Textiles 19 (3)*:73 (1990).
18. Gelman Sciences Technology Ltd., Class 10 for clean rooms, *Textile Horizons 7* (5):12 (1987).
19. C. E. Wood, Repellency and breathability in medical nonwoven fabrics, *J. Coated Fabrics 19* (1):143 (1990).
20. K. Slater, Greater comfort for surgeons, *Medical Textiles 4* (6):1 (1987).
21. Occupational Safety and Health Administration, Occupational Exposure to Bloodborne pathogens; Final Rule, *Fed. Reg. 56* (235):64004 (Dec. 6, 1991).
22. Anon., Getting ready to tape the bugs, *Chem. Week 115* (8):43 (1974).
23. C. D. Potter, Porous acrylic fibres for controlled release applications, *J. Coated Fabrics 18* (2):259 (1989).
24. T. Vigo, Antimicrobial fibers and polymers, *Manmade Fibers: Their Origin and Development*, (R. B. Seymour and R. S. Porter, eds), Elsevier Applied Science Publishers, London, 1993, p. 214.
25. D. D. Gagliardi, Antibacterial finishes, *Am. Dyest. Reptr. 51* (2): P49 (1961).
26. L. S. Vol'f, A. I. Meos, V. V. Koteskii, and S. Hillers, Biologically active poly(vinyl alcohol) fibers, *Khim. Volonka* (6):16 (1963).
27. G. F. Danna, T. L. Vigo, and C. M. Welch, Permox-a hydrogen peroxide-zinc acetate antibacterial agent for cotton, *Text. Res. J. 48*:173 (1978).
28. A. J. Isquith, E. A. Abbott, and P. A. Walters, Surface-bonded antimicrobial activity of an organosilicon quaternary ammonium chloride, *Appl. Microbiol. 24* (6):859 (1972).
29. Z. Hagiwara, O. Hideo, H. Shigetaka, N. Saburo, S. Ida, and T. Kenichi, Particle-packed fiber article having antibacterial property, *Eur. Pat. Appl. EP 103,214* (Mar. 21, 1984).
30. R. H. McIntosh, Jr., A. F. Turbak, and R. H. McIntosh, Sr., Biocidal delivery system of phosphate ester and method of preparation therof, *U.S. Patent 4,908,209* (Mar. 13. 1990).
31. P. E. Hoch, G. M. Wagner, and W. J. Vullo, The bactericidal properties of THPC-resinated cotton fabric, *Text. Res. J. 36* (8):757 (1966).
32. Kanebo Ltd., Cleaner garments for clean rooms, *Medical Textiles 8* (1):5 (1991).
33. T. L. Vigo, G. E. R. Lamb, S. Kepka, and B. Miller, Abrasion and lint loss properties of fabrics containing crosslinked polyethylene glycol, *Text. Res. J. 60* (3):169 (1990).
34. T. L. Vigo and J. S. Bruno, Applications of phase change polymers in fibrous substrates, *Proc. 26th Intersoc. Energy Conversion and Eng. Conf. Vol. 4*, 1991, p. 161, Boston, Ma., Aug. 4–9, 1991.
35. Canadian Ministry of Defence, B. Farnworth, and J. K. Dix, Skin-tight chemical/biological protective suit, U.S. Patent 5,017,424 (May 21, 1991).

36. R. W. Sidwell, G. J. Dixon, L. Westbrook, and F. H. Forziati, Quantitative studies on fabrics as disseminators of viruses. IV. Virus transmission by dry contact of fabrics, *Appl. Microbiol. 19* (6):950 (1969).
37. R. W. Sidwell, G. J. Dixon, L. Westbrook, and F. H. Forziati, Quantitative studies on fabrics as disseminators of viruses. V. Effect of laundering on poliovirus-contaminated fabrics, *Appl. Microbiol. 21* (2):227 (1971).
38. J. N. Mbithi, V. S. Springthorpe, and S. A. Syed, Chemical disinfection of hepatitis A virus on environmental surfaces, *Appl. Environ. Microbiol. 56* (11):3601 (1990).
39. Allied Fibres, Lightweight gloves offer extra protection. *Medical Textiles 7* (11):6 (1990).
40. R. Baker, Consumer awareness and healthcare concerns promote interest in antimicrobial-treated carpets, *Text. Chem. Color. 21* (8):10 (1989).

10

Biological Test Methods for Protective Clothing

NORMAN W. HENRY III E. I. du Pont de Nemours & Company, Newark, Delaware

I. INTRODUCTION

The growing concern of health care professionals about potential infection from human immunoviruses (HIVs) and hepatitis B virus (HBV) has stimulated renewed interest in developing biological test methods for clothing. Biological hazards pose an interesting challenge for test method development because microbial pathogens can be transferred either in liquids or in air as aerosols. Exposed body surfaces and broken skin are vulnerable to these pathogens unless an effective barrier of clothing is worn for protection. This chapter will share some of the developments of biological test methods for determining clothing resistance to microbiological hazards.

The father of microbiology, Louis Pasteur, was first to demonstrate the possible mode of transmission of microorganisms in air. Using a round bottom flask with a long open ended neck (retort) containing sterilized nutrient broth, he showed that it took a much longer time for microbial growth to occur in this flask compared to an open wide mouth flask with the same sterilized broth and no neck exposed to room air. The long narrow glass neck provided a long torturous path for the microorganisms to pass through before reaching the nutrient broth. Likewise, clothing provides the same torturous path for preventing exposure to airborne pathogens. This path, however, can be altered by the effect of liquids, such as blood, that act as carriers causing clothing to shrink or swell upon contact and allowing penetration, thus providing a continuous flow of microorganisms through clothing materials.

Blood and other body liquids all possess distinctive physical and chemical properties, such as surface tension, that effect adsorption on the surface and through clothing materials. Penetration of liquids containing bacteria and virus particles of different sizes and shapes is best described as the non-molecular movement of microorganisms through barrier clothing materials. Hence, because clothing interacts with liquids, biological test methods must be designed to determine penetration resistance of microorganisms at exposure conditions with liquids that actually contain some surrogate organism that can be detected by a microbiological technique.

Because clothing is the last line of defense from exposure to bloodborne pathogens in the health care industry, there never has been a greater need for a standardized biological test method. Widespread concern about infection and disease from these pathogens has caused health care professionals to question the performance of their clothing. Surgeons, doctors, nurses, paramedics, and emergency response personnel are just as concerned about patients contaminating them as they are concerned about contaminating their patients. There are over 5.6 million health care workers today, and of these workers the Centers for Disease Control and Prevention (CDC) reported that there are approximately 40 infected with AIDS because they came in contact with body liquids of persons infected with the disease or actually worked with and were accidentally exposed to the live virus in research activities. Also it has been estimated that there are between 6,000–7,500 cases of HBV infection each year. In response to this problem, the Occupational Safety and Health Administration (OSHA) recently published its final bloodborne standard on December 6, 1991. This standard requires that those individuals in health care fields have an exposure control plan in place that includes wearing appropriate clothing and gloves when drawing blood or having potential hand contact with blood and other body liquids.

II. BACKGROUND

Clothing is worn to protect against physical and chemical exposures that may cause harm in the chemical manufacturing industry. The importance of clothing for protecting against biological exposures to microorganisms was well recognized in the medical profession where surgical gloves and gowns were first worn to prevent infecting patients and their wounds from bacterial contamination during surgery. Reports in the literature describe the first use of rubber gloves in surgical procedures at Johns Hopkins University Medical School back in the early 1900s [1]. Surgical gloves and gowns were first sterilized and then worn by surgeons to protect the patient from exposure.

Now, as before, gloves and gowns are worn not only to protect the patient but also to protect the surgeon or other health care professionals and their clothes from contamination with patient's pathogens. Blood containing these pathogens and aerosols produced during surgical and laboratory procedures are prevented from contacting the body by a variety of items of clothing that act as a barrier.

There are a number of different types of protective clothing used today. Some of these are disposable and nondisposable. These items include caps, gloves, gowns, suits, and booties. They all are constructed from materials that are woven, nonwoven, and laminates with unique properties that are intended to afford protection from liquid contact with microbial pathogens. At the same time, they are expected to be barrierproof and breathable in order to provide comfort to the worker. Similar types of laboratory clothing and gloves are also worn by research personnel working with infectious agents under controlled conditions in research institutions, hospitals, and universities.

III. METHODS

There are a number of different types of test methods for determining whether clothing is liquid-resistant. To date, there still are no standard test methods that are universally accepted, although standard setting organizations such as the American Society for Testing Materials' (ASTM) Committee F-23 on Protective Clothing is working on draft methods. The American Association of Textile Chemists and Colorists (AATCC) has several water-resistant methods and (INDA), an association of the nonwovens fabrics industry, has the Mason Jar test. Additional tests have been proposed by manufacturers, but attempts to standardize them in organizations such as the Association for the Advancement of Medical Instrumentation (AAMI) were unsuccessful. A summary of some of the current test methods available is shown in Table 1. In general, the test methods can be separated into those for gloves and those for gown or suit materials. These methods range from simple leak tests for pinholes and imperfections to more complex microbiological tests with various microorganisms.

One of the earliest and most widely used test methods for determining liquid penetration through clothing materials was the mason jar test. This test consisted of putting a sample of clothing material over the top of a mason jar filled with saline or any other liquid, inverting it so that water came in contact with the outside surface of the clothing material and observing for liquid penetration and surface wetting. This test was designed to use a 114-mm hydrostatic head or 600-mL of saline solution to challenge the clothing. It was

Table 1 Summary of Some Biological Test Methods for Items of Clothing

Gloves	Gowns, coveralls, and suits
Mason Jar (INDA)	Mason Jar (INDA)
Electorconductive (DOD)	Impact Penetration Test (AATCC)
Colorimetric	Hydrostatic Pressure Test (AATCC)
Air Inflation (ASTM)	Synthetic Blood (ASTM F23.40.01)
	Bacteriophage (ASTM F23.40.02)
	Kimberly-Clark Strike-Through
	Gore Elbow Lean Test
	Surface Penetration Test
	Liquid Penetration Stress (Hammock Test)

developed by the Medical/Surgical committee of INDA. No time limits were given for the test; although it was easy and economical to set up and operate, it was not recommended because it tested water penetration rather than microbial penetration. Modified versions of the Mason Jar test were investigated to determine whether liquid penetration was required for microbial transfer. After clothing samples were exposed to bacterial solution on their outside surface, the inside surface of the samples were contacted and innoculated in Petri dishes to see if the organisms grew if no liquid penetrated. Results showed that the organisms did not grow if no liquid penetrated. Similar tests were done with gloves filled with a culture of bacteria so that the fingers dangled within a sterile culture medium. If there was a small pinhole or leak, the bacteria passed through into the sterile media and began to grow.

Another innovative testing device that was developed to detect holes in gloves during operations was reported by William C. Beck [2]. The device was set up by Edward Weck and Company and called the *Wecktester*. It was set up in an operating room having a conductive floor, a battery, a basin of saline solution, and a monitoring device. The surgeon would check his gloves for leaks during an operation by dipping them in the basin of saline while standing on the conductive floor. If the gloves leaked, the monitoring device would show a deflection and the surgeon would immediately discard the glove. Although this test gave immediate results in the operating room, it was not practical for other operating rooms in the country without conductive floors.

Several other glove test methods should also be noted. The first is an electroconductance method for detecting defects in disposable latex and plastic gloves prescribed by the Department of Defense (DOD). In this test gloves to be tested are filled with physiological saline solution and placed in a stainless steel beaker also containing saline solution. A cylindrical copper tube placed in the glove serves as one electrode. The presence of holes is demonstrated by a current flow through the circuit as noted by a deflection in of the needle of a milliammeter. A colorimetric method also has been used. In this test method, gloves to be tested are filled with 300 mL of a saturated solution of methylene blue solution and lowered into a beaker of water. The presence of holes is demonstrated by the appearance of the indicator in the water in the beaker. Both of these methods were reported by Ballbach et al. [3].

ASTM's Committee D-11 on Rubber also devised a standard test method for rubber surgical gloves, D 3577 -78a (Reapproved 1982). This standard specified certain requirements for packaged sterile rubber gloves. In this test, holes in gloves were detected by fastening the cuff to a circular mandrel, inflating with air to a gage pressure of 1.5 kPa and immersing in water at room temperature to a depth of 200 mm above the tip of the middle finger. The inflation pressure was 10%, and the immersion time, 1.5 minutes. Emergence of air bubbles from the glove constituted an air-tight failure. A picture of this test apparatus is shown in Figure 1.

Other water-resistant test methods for clothing published by the AATCC are the Impact Penetration test (AATCC 42-1985) and the Hydrostatic Pressure test (AATCC 127-1985). These methods are primarily used for textile fabrics. The impact penetration test is used to determine the penetration of liquids upon impact (e.g., splash by a liquid during a surgical procedure). In this test, an AATCC Impact Penetration Tester is used with blotter paper. The blotter paper is weighed on an analytical balance and placed under the surface of sample of clothing material. The sample is sprayed on its outside surface with 500 mL of liquid from a height of 61 cm. The blotter paper is then reweighed after exposure to liquid impact. The increase in weight is reported as the impact resistance.

The hydrostatic pressure test measures the penetration of liquids under steadily increasing pressure. A Hydrostatic Pressure Tester is used for this test. The test consists of mounting a test sample of clothing under an oriface of a conical well in the tester and subjecting it to water pressure increasing at a constant rate until three points of leakage appear on its under surface. The hydrostatic head pressure at the moment of penetration measured in centimeters is reported.

The latest developments in biological clothing test methods are standards drafted by ASTM's Committee F-23 on Protective Clothing. Subcommittee

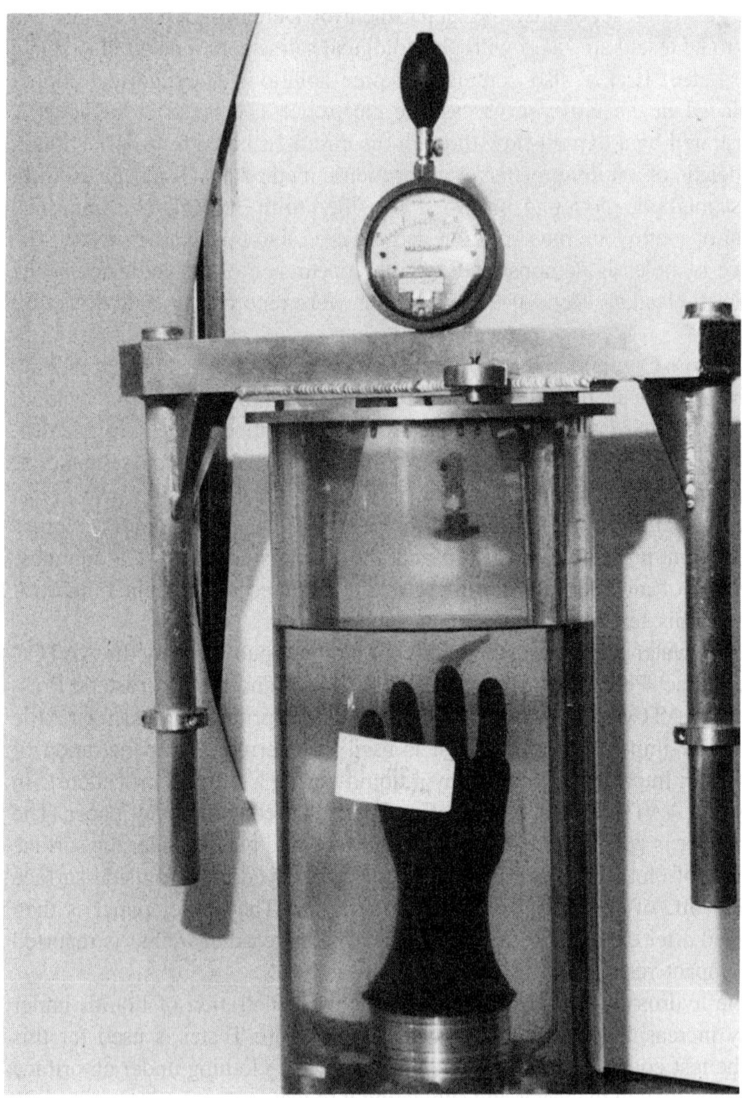

Figure 1 Test apparatus used for measuring water resistance in sterile rubber gloves.

F23.40 on biological test methods believes that the two methods they have proposed will provide initial performance data to select clothing for biological resistance, just as the chemical permeation test method (ASTM F739-92) provided users chemical resistance on specific chemicals. Of the two methods, F23.40.01 would provide a screen for eliminating those clothing materials that allowed synthetic blood to penetrate through, before going on to the more extensive bacteriophage test method F23.40.02. In this test method a bacteriophage (Phi-X174) is used as a surrogate microbe for HBV and HIVs in a suspension at a titer of 10^8 under the condition of continuous contact with clothing material for 1 hr, at zero pressure for 5 min, 2 psi for 1 min and 0 psi for the remaining minutes or until visible penetration is observed. Visual detection is supplemented with an assay procedure that can detect viable virus that may penetrate the clothing material even when liquid or blood penetration is not visible. Both of these test methods are pass/-fail tests that use challenge media (synthetic blood and a surrogate virus) and test conditions that simulate potential exposure conditions. Although they are not representative of all exposures, they do provide an initial evaluation for making a selection. Furthermore, both of these methods have undergone successful interlaboratory round robin tests with participating research, government, and manufacturers' laboratories. A picture of the penetration test apparatus is shown in Figure 2.

Acceptance of these methods for standards has been a difficult process for F23.40. Existing test methods such as the Mason Jar and hydrostatic head pressure method universally used by all manufacturers were considered to be the best practical methods available. They were well established, did not require new equipment or test apparatus, and were inexpensive. However, these methods used water or saline as a challenge liquid and were run at different pressures and test conditions. Also, they would not be indicative of viral penetration. Other test methods have been proposed by various manufacturers using either bovine or synthetic blood. One method, the Kimberly-Clark Strike-Through method, involves putting droplets of bovine blood on the outside surface of the clothing material, placing a preweighed AATCC filter paper underneath on the inside surface, and pressurizing the whole system (clothing and filter paper) until visible strike-through is observed on the filter paper. By weighing the filter paper after exposure, one can quantitate percent strike-through. This method is still being evaluated by F23.40 with interlaboratory tests. Another test method that dramatically shows penetration is an elbow lean test developed by Gore and Associates. In this test, an item of clothing is placed against a presoaked blotter containing synthetic blood, and a piece of parafilm is placed over the inside clothing surface. Users

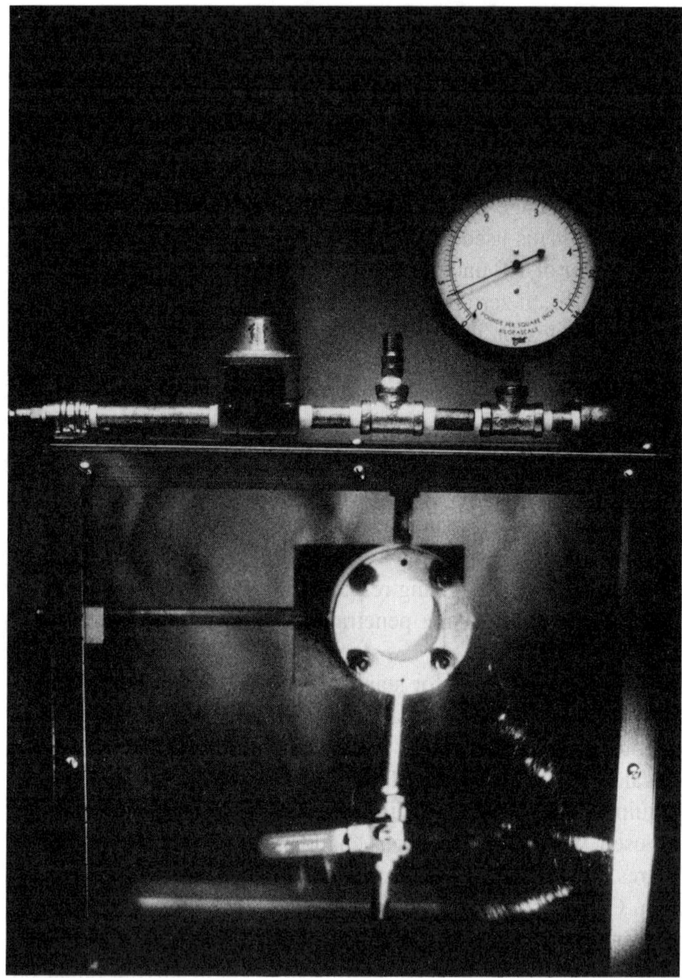

Figure 2 ASTM F903-90 penetration test apparatus used for measuring synthetic blood and microbiological penetration.

conducting the test merely lean their elbows on the parafilm cover until visible penetration is observed. Ergonomic measurements of the pressure used in this test approximate the 2-psi pressure recommended in ASTM's draft methods with synthetic blood and the Phi X174 surrogate microbe. Both Kimberly Clark's strike-through and Gore's elbow lean test are currently being considered as alternative tests by subcommittee F23.40.

Table 2 Organizations Involved in Developing Test Methods

Association of the Nonwoven Fabrics Industry (INDA)
American Society for Testing and Materials (ASTM)
American Association of Textile Chemists and Colorists (AATCC)
Association for the Advancement of Medical Instrumentation (AAMI)
Association of Operating Room Nurses (AORN)

IV. SUMMARY AND CONCLUSION

A number of different biological test methods for determining the resistance of clothing to biological hazards have been presented. They range from simple leak tests to more extensive complex tests with synthetic blood and surrogate microorganisms at various pressures and test conditions. No universal test apparatus can be used for all the tests, and there are no universal test conditions that represent all potential exposures. A comparative study of some of these test methods with nine surgical gown fabrics has been reported by McCullough et al. [4]. Some correlation was seen between the methods. Results indicated that synthetic blood is recommended for barrier testing and that 2 psi should be the recommended pressure for use. Additional methods may be needed, but at least for now it appears that ASTM's synthetic blood and bacteriophage tests (F23.40.01 and F23.40.02, respectively) show promise of being accepted as consensus standards.

This summary of biological test methods is far from complete. Table 2 lists some organizations currently involved with developing test methods. Additional references to articles on biological resistance test methods are listed at the end of this chapter. These methods and studies represent an attempt by end users in hospitals, manufacturers, and regulators to evaluate the performance of clothing to biological liquids and organisms that are responsible for disease and infection. Hopefully, the need for performance standards will stimulate the development of improved protective clothing and reduce the the incidence of disease. More than one standard test method may be needed, but at least for now the methods provide data for selecting effective barrier materials that reduce the risk of exposure.

REFERENCES

1. W. S. Halstead, Ligature and suture material: Also an account of the introduction of gloves, *JAMA*, 1119 (1913).
2. W. Beck, Holes in rubber gloves description of a new instrument to detect holes during operations; *Guthrie Clin. Bull. 29*:14, (1959).

3. Ballbach et al., A study of testing methods for the detection of defects in disposable latex and plastic gloves, *Journal of the AOAC* (1):1074 (1972).
4. McCullough et al., Liquid barrier properties of nine surgical gown fabrics, *INDA Journal of Nonwovens Research* (3) (Summer 1991).

BIBLIOGRAPHY

1. Laufman et al., Scanning electron microscopy of moist bacterial strike-through of surgical materials, *Surgery, Gynecology & Obstetrics 150*:165 (February 1980).
2. Schwartz et al., Microbial penetration of surgical gown materials, *Surgery, Gynecology & Obstetrics 150*:507 (April 1980).
3. Reeves, Fluid penetration testing for medical protective garb, *Tappi Journal*, (November 1990).
4. Smith, Barrier efficiency of surgical gowns, are we really protected from our patient's pathogens?, *Archives of Surgery 126*:756 (1991).
5. Carey et al., A laboratory evaluation of standard leak tests for surgical & examination gloves, *Journal of Clinical Engineering 14* (2) (1989).
6. N. Henry III and D. C. Montefiori, The resistance of clothing materials to biological liquids, *Performance of Protective Clothing*, ASTM STP 1133 (J. P. Mc Briarty and N. W. Henry III, eds.), American Society for Testing and Materials, Philadelphia, 1991.

Index